S0-EXP-521

New Perspectives in Wood Anatomy

Forestry Sciences

Also in this series:

Oldeman RAA, et al, eds: Tropical Hardwood Utilization: Practice and Prospects. 1982. ISBN 90-247-2581-x

Prins CFL, ed: Production, Marketing and Use of Finger-Jointed Sawnwood. 1982. ISBN 90-247-2569-0

In preparation:

Bonga JM and Durzan DJ, eds: Tissue Culture in Forestry. 1982. ISBN 90-247-2660-3

Chandler CC, Cheney P and Williams DF, eds: Fire in Forest

Den Ouden P and Boom BK, eds: Manual of Cultivated Conifers: hardy in cold- and warm-temperate zone. 1982. ISBN 90-247-2148-2 paperback; ISBN 90-247-2644-1 hardbound

Gordon JC and Wheeler CT, eds: Biological Nitrogen Fixation in Forest Ecosystems: Foundation and Applications

Hummel FC, ed: Forestry Policy

Satoo T and Madgwick HAI: Forest Biomass. 1982. ISBN 90-247-2710-3

Németh MV: The Virus-Mycoplasma and Rickettsia Disease of Fruit Trees

Powers RF and Miller HG, eds: Applied Aspects of Forest Tree Nutrition

Powers RF and Miller HG, eds: Basic Aspects of Forest Tree Nutrition

Rajagopal R: Information Analysis for Resource Management

Van Nao T: Forest Fire Prevention and Control

New Perspectives in Wood Anatomy

Published on the occasion of the 50th Anniversary of the
International Association of Wood Anatomists

Edited by
P. Baas
Rijksherbarium, Leiden, The Netherlands

1982

MARTINUS NIJHOFF/DR W. JUNK PUBLISHERS
THE HAGUE/BOSTON/LONDON

IV

Distributors:

for the United States and Canada
Kluwer Boston, Inc.
190 Old Derby Street
Hingham, MA 02043
USA

for all other countries
Kluwer Academic Publishers Group
Distribution Center
P.O.Box 322
3300 AH Dordrecht
The Netherlands

Library of Congress Cataloging in Publication Data
Main entry under title:

New perspectives in wood anatomy.

    (Forestry sciences
    "Published on the occasion of the 50th
anniversary of the International Association of
Wood Anatomists."
    Includes index.
    1. Wood--Anatomy--Addresses, essays, lectures.
2. International Association of Wood Anatomists--
Addresses, essays, lectures. I. Baas, P.
II. International Association of Wood Anatomists.
III. Series.
QK647.N48   582'.041   82-2281

ISBN 90-247-2526-7   AACR2

ISBN 90-247-2526-7 (this volume)
ISBN 90-247-2447-3 (series)

PRINTED IN THE NETHERLANDS

Contents

Preface

On the occasion of the 50th Anniversary of the *International Association of Wood Anatomists* several symposia were held during the 13th International Botanical Congress in Sydney, August 1981. Extended versions of most of the invited papers presented there, and some additional papers on aspects which could not be included in the congress program constitute the contents of this book, which intentionally received the pretentious title *'New Perspectives in Wood Anatomy'*.

To some readers it may seem a paradox that under this heading papers on a diversity of partly traditional wood anatomical subjects are assembled, even including two with a historical emphasis. However, a study of the history of wood anatomy and of how students of that discipline joined forces in an international association, brings to light many facts and views which deserve the attention of present day and future wood scientists as a potential source of inspiration for their research and organisational work.

In this volume equal attention is paid to pure and applied aspects of wood anatomy. The significance of wood anatomy for the functioning of trees as well as for the performance of timber, the development of wood in the tree and the possibilities of genetic manipulation and improvement, together with the story wood anatomy has to tell about the evolution of woody plants are disussed, as well as the practically important, new developments in applying statistics and computers to wood anatomical research. In all these fields there are promising new perspectives, which receive full attention from the authors, and which deserve to be actively pursued in the research of the coming decades. The perspectives will be all the more promising if the various specialisations brought together here are viewed against the background of wood anatomy as an integrated discipline, in which these seemingly diverging specialisations converge again to contribute to our general understanding of wood.

The editing of this book has been a light task thanks to cooperation of all the authors, and above all, thanks to the invaluable assistance of Emmy van Nieuwkoop in preparing the manuscripts for the press. I also wish to acknowledge the support from the staff of Martinus Nijhoff/Dr. W. Junk Publishers in the emergence of this book. I hope it will be of value to all with an interest in the structure and properties of one of nature's most pleasing resources: wood.

Leiden, Rijksherbarium, 1982
P. Baas

Highlights in the early history of the International Association of Wood Anatomists

WILLIAM LOUIS STERN

Department of Botany, University of Florida, Gainesville, Florida 32611, U.S.A.

Summary: The vision of Laurence Chalk (Imperial Forestry Institute, Oxford), B.J. Rendle (Forest Products Research Laboratory, Princes Risborough), and Samuel J. Record (Yale School of Forestry) led to meetings among wood anatomists at the Fifth International Botanical Congress, Cambridge, England during August 1930 for the purpose of holding 'an informal conference on Systematic Wood Anatomy.' Following these discussions a committee was formed to consider the question of a formal organisation and to report at the next conference. This was provisionally fixed for July 1931 in Paris coincident with meetings of the Congrès International du Bois et de la Sylviculture. At these meetings a constitution was drafted and an Organizing Committee of 9 was established with Record as secretary and Rendle as deputy. This group was charged with enrolling members and carrying out the affairs of the infant Association until a formal Council, as provided in the constitution, could be elected. To this end Record invited a group of internationally known wood anatomists to join the Association as founder or charter members. Thirty-six persons from 14 countries signified their intent to associate with the new society and to participate in its activities. Among the latter were the interchange of ideas and information on wood anatomy, development of a standard terminology, collection and exchange of wood specimens, and encouragement of teaching, research, and publication on wood structure. Members were formally nominated and admission to membership was controlled by the Council. Since these early days the Association has broadened its activities and objectives in conjunction with new technology and expanded fields of anatomical interest. Membership is open, without formal vote, to any actively participating in the aims of the Association.

The concept of some kind of international organisation of wood anatomists was already in the mind of Samuel James Record when he visited Europe in 1928. In an interview with T.F. Chipp at the Royal Botanic Gardens, Kew, he inquired about the possibility of holding an informal gathering of wood anatomists during the Fifth International Botanical Congress set to convene at

Cambridge in 1930. Record was to leave that meeting knowing there would be no objection to his arranging an unofficial meeting. He communicated then (in 1929; *fide* B.J. Rendle, personal correspondence) with Laurence Chalk and Bernard J. Rendle (neither of whom he had yet met) to seek their interest and help and together they agreed to work toward international cooperation for the betterment of their science and the application of that science to the solution of practical problems. Even though Record boasted to Chalk in 1934 that the three of them had 'just cause for pride in the way the idea of a world-wide cooperation has been working out [and that] the future [for wood anatomy] is going to be infinitely better than in the past,' this bravado was not always characteristic of Record's predictions for the success of the assembly of wood anatomists which was to grow into the International Association of Wood Anatomists. International cooperation in science, like world coopera-tion in economics and politics, does not spring full-blown from the drawing board. Achievement of the anatomical nirvana envisioned by Record was not to be realised without infinite endurance, heartbreaking disappointments and frustrations, and letter after letter of sincere, constructive, and sometimes very personal criticism among the three principal architects of the Internation-al Association of Wood Anatomists. That they persevered in their aims in the face of a distant group of colleagues, some of whom were only heard from when vested interests were infringed or pet viewpoints challenged, is testimony to their doggedness and singleminded belief in the contributions wood anatomy could make to science through the collective efforts of specialists bound together by a common cause.

In 1928, Laurence Chalk was Lecturer in the Structure and Properties of Wood in the Imperial (now Commonwealth) Forestry Institute, Oxford, and was working for his doctorate. He had already seen service with the Forest Department in Uganda where, probably, he first gained an interest in tree growth, tropical woods, and wood structure. Apart from his time in Uganda Chalk's entire professional career was spent at Oxford University where he eventually became University Reader in Wood Anatomy and a Lecturer of Uni-versity College. During his early days on the Oxford faculty, he met Bernard J. Rendle who worked with him in the same building at Oxford pending the con-struction of the Forest Products Research Laboratory at Princes Risborough (now, Princes Risborough Laboratory, Building Research Establishment).

Bernard J. Rendle was head of the Wood Structure Section (now, Proper-ties of Materials) of the Forest Products Research Laboratory and continued in that capacity until his retirement. The laboratory is some 20 miles from Oxford and long after Rendle left the University to move into the newly con-structed buildings at Princes Risborough, he kept in close touch with Chalk.

The indefatigable first Secretary-Treasurer of the International Association of Wood Anatomists, Samuel James Record, was Professor of Forest Products in the School of Forestry, Yale University, during the cradle days of the Association. By 1928 he had already published three books on woods, established and edited the journal *Tropical Woods*, and brought together a collection of wood specimens which was to become pre-eminent in the world. Record had a fairly free hand to carry out his aims at international cooperation in wood anatomy in contrast with his soon-to-be-well-known contemporary, Arthur Koehler, 'who was debarred by his official position and duties [at the U.S. Forest Products Laboratory] from taking an active part in an international organization - or so he said' (B.J. Rendle, personal correspondence). His correspondence was cosmopolitan and it can be said fairly, he was to become Nestor of wood anatomists.

Thus stand the three designers of the International Association of Wood Anatomists: An academician affiliated with the venerable Oxford University and the resources of the associated Imperial Forestry Institute who had an abiding knowledge of the growth, formation, and structure of wood; a scientific civil servant confronted with the practical day-to-day problems of wood identification, properties, names, and uses of timbers; and a faculty member of the prestigious School of Forestry at Yale, founder and editor of a journal devoted to tropical forests and tropical woods, specialist in comparative and systematic wood anatomy, and curator of the largest wood collection then extant. Together, these men comprised a formidable triumvirate capable of sorting out the intricacies of wood anatomy, on the one hand, and the personal strengths and foibles of wood anatomists, on the other.

Early records indicate that among Chalk, Record, and Rendle their major concern with respect to international cooperation in wood anatomy was aimed at the standardisation of terminology. They agreed that the best way to begin the process of standardisation was for those interested in the matter to meet informally during the Fifth International Botanical Congress which was to be held at Cambridge, England in August, 1930. Accordingly, in February, 1930, a printed letter was circulated from the Imperial Forestry Institute at Oxford inviting all persons interested in the systematic anatomy of woods to meet together at Cambridge; the invitation was also published in *Tropical Woods* in June of that same year. Chalk, Record, and Rendle signed the letter and article as conveners of the conference.

The goals of the two Englishmen did not appear at that time very ambitious and, as Chalk stated in a letter to Record of March, 1930, 'Neither Rendle nor I are expecting to get very much done in the way of solving particular points of nomenclature ... and are in favour of concentrating on an effort to secure

4

Fig. 1. Samuel James Record. – Fig. 2. Laurence Chalk (left) with Adriance S. Foster, 1964. – Fig. 3. Bernard J. Rendle, between 1977 and 1979. – Fig. 4. Participants at the August 20, 1930 Cambridge meetings. Identifications are from Record's personal copy of the original photograph. From left to right, standing: H.S. Holden, Nottingham, E.H.B. Boulton, Cambridge, Jean Collardet, Paris, F.W. Jane, London, W. Dawson, Cambridge, B.J. Rendle, Princes Risborough, Samuel James Record, New Haven, Laurence Chalk, Oxford, Paul Ledoux, Brussels, Charles Russell Metcalfe, Kew; sitting: Llewelyn Williams, Chicago, S.H. Clarke, Princes Risborough, S.E. Wilson, England.

agreement to some plan for leaving the work in the hands of a small committee, possibly with the idea of presenting the results of such labours to the following Congress for its blessing.' But, reading through the collected correspondence of the three Fathers of the I.A.W.A., this was probably not the way Record viewed the potential outcome of that meeting at all, for he had been, on his own, preparing a polyglot glossary of terms in wood anatomy and doubtless wished the international sanction for his work which might come from just such an assembly. Record, it will be seen, had been writing his 'Identification of the Timbers of North America,' and had, accordingly, been using just such descriptive terms as were then under question. Be this as it may, Chalk allowed that 'In the majority of cases there is no actual confusion as to the meaning of terms, and it is merely a question of deciding to stick to one out of two or three synonymous expressions.' Thus, the call for the meeting in 1930 at Cambridge stressed 'the possibility of introducing some measure of standardization [and also to] arrange for a scheme for the exchange of material among botanists and forest botanists who are willing to undertake the study of a family or group.' The conveners hoped for cooperation and an exchange of views to take place during the Cambridge conference.

At Cambridge the first meeting was held in the building of the Department of Forestry where the three conveners explained the purpose of the conference more fully than was possible in the letter of invitation. There were suggestions that the purpose of the organisation should include the entire spectrum of forest products research, but a majority held that the scope should be limited to the original proposal, that is, to wood anatomy research. A committee was appointed to draft a constitution for presentation and consideration the following day.

Attempts to draft a constitution and establish an organisational structure were immediately fraught with a host of differing opinions concerning eligibility for membership and suggestions for different grades of membership each to be affiliated in some manner or other. On top of this, it developed that the proposed form of constitution did not fully meet certain vague international legal requirements. It proved later on that this objection was wholly inconsequential and may have been interjected as a straw man to delay immediate establishment of a constitution without wider discussions. A major highlight and outcome of these first organisational meetings was the name, International Association of Wood Anatomists, adopted for the yet-to-be-established society.

To strengthen the momentum toward organisation, at the third meeting on August 20, it was recommended that an Organizing Committee be appointed with full powers 'to deal with the question and report to the next conference, to be held in Paris next summer on the occasion of the Congrès International des Bois Tropicaux et Subtropicaux [actually, Congrès International du Bois et de la Sylviculture].' The recommendation was adopted and a committee of nine members elected. There was every attempt to make this committee as internationally representative as possible, but of necessity, membership had to be limited to those in attendance at Cambridge and to those who had indicated by letter their willingness to cooperate. Membership on that committee set the pattern for the future of the International Association of Wood Anatomists, and its structure, therefore, should be recorded here:

> E.H. Brooke Boulton, Department of Forestry, University of Cambridge, England.
> G. Bredemann, Staatsinstitut für angewandte Botanik, Hamburg, Germany.
> Laurence Chalk, Imperial Forestry Institute, University of Oxford, England.
> Jean Collardet, Comité National des Bois Coloniaux, Paris, France.
> Paul Ledoux, Institut Botanique Léo Errara, Brussels, Belgium.
> J.H. Pfeiffer, Technische Hogeschool, Delft, The Netherlands.
> Samuel J. Record, School of Forestry, Yale University, New Haven, U.S.A.
> Bernard J. Rendle, Forest Products Research Laboratory, Princes Risborough, England.
> M.B. Welch, Technological Museum, Sydney, Australia.

On August 21, those members of the Organizing Committee present at Cambridge - Boulton, Chalk, Collardet, Ledoux, Record, and Rendle - together with E. Reinders, representing the Netherlands and Java, met to begin planning for the next conference to be held in Paris. They appointed Record Secretary of the Committee and established that the Committee expected to present 'at least' three reports the following year: 1) a definite plan for organisation (*i.e.*, a constitution), 2) a plan for the exchange of material (*i.e.*, wood specimens), and 3) a polyglot glossary of terms used in describing woods! It is surprising that a committee comprising scientists of wide experience could hope to carry out such an ambitious program within the space of one year, especially as they would have to depend upon the prompt and full cooperation of a group of far-flung committee members, some of whom were as yet unaware of their appointment, and other concerned wood anatomists in distant corners of the world. That they succeeded as well as they did, though, is tribute to the persistence, willingness to do most of the work, and frustration-absorbing qualities of Chalk, Record, and Rendle, who, after all, were the prime initiators of the Association.

Record returned from Cambridge to the United States on September 22,

1930 but owing to the press of accumulated material and routine business at Yale was unable to communicate with the Organizing Committee until October 15. However, there followed a spate of circular letters advising the members of his activities as Secretary and imploring their assistance and cooperation in various matters pertaining to the formal establishment of the I.A.W.A. At the same time, Record continued his active correspondence with Chalk and Rendle and wrote dozens of letters to prospective I.A.W.A. members, both to individuals and institutions throughout the world eliciting their favourable interest. Meanwhile, he continued to edit *Tropical Woods*, work on his 'Identification of the Timbers of North America,' write articles and monographs on tropical woods and trees, attend to the sundry duties of the curator of a major collection of wood specimens and a university Professor of Forest Products. Reading the litany of Record's day-to-day activities reminds one of Linnaeus's recitation of his labours which end with his prayer, 'If the Almighty grants me a few more years I will release the aging horse from the yoke so that he will not entirely collapse and end up being a laughing stock. If then I succeed in having my garden and a few rare plants, I shall rejoice in them.'

There were communication problems right at the outset and by March, 1931 Record had despaired of ever hearing from either Ledoux or Boulton concerning answers to questions put to them about the organisation of the Association. He wrote bitterly to Chalk that month, 'so far as I am concerned, Ledoux and Boulton are both dead until they convince me otherwise. I did not send them copies of the glossary [for review] and shall waste no more postage upon them until they come to life.' Later correspondence shows, however, that Ledoux and Boulton were duly engaged in legitimate business which prevented their prompt responses to Record's insistent queries. But it is easy to understand the exasperation of such an intense individual as Record over the apparent recalcitrance of colleagues who had agreed to cooperate with him in an important enterprise especially as he was charged with the major responsibility to assure its success.

Record had some rather firm ideas about potential membership of the Association and conveyed these to his friends Chalk and Rendle. It appears Record felt election to Association membership ought to be limited and that only those scientists who had demonstrated their ability to conduct research in wood anatomy should be eligible for full membership. There was disagreement with Chalk and Rendle about classes of membership. In November, 1930, Chalk wrote to Record in this respect to indicate that 'Rendle and I are not keen on 'Fellows' mainly because of the small size of the show. There would be very few qualified to be members who would not also be eligible for

8

fellowship. I do not think a great deal would be gained by the distinction ....
and it might discourage some people. Incidentally fellowship of an association
would sound very odd in this country. The 'Associates' we agree about. With
regard to 'Patrons', living in a more democratic country than yours, I am averse
to patronage in principle, but there seemed to be a strong feeling amongst the
continental representatives in favour of such a grade. ... Our views therefore
are that the grades should be 1. Members (The only grade entitled to vote). 2.
Associates. 3. Patrons.' Record replied: 'I agree with you and Rendle in the
matter of leaving out the 'Fellows'. My object in suggesting it was to find some
way to restrict the important membership of the Association to persons who are
prominent in the field of wood anatomy. I should like to see it made what we
call a 'professional' society and have some standing in the world.'

In a progress report to the Organizing Committee of February, 1931, Record
outlined achievements to date: 'It appears to me that our accomplishment
consists almost entirely of propaganda. We have been fairly successful in
arousing widespread and favorable interest in the cause we represent.' He goes
on to reiterate for the Committee's benefit the distribution of responsibilities
among them agreed upon at Cambridge. Among these were that Collardet be
delegated to organise the Paris meeting during the summer of 1931; that
Collardet and Ledoux be delegated to report on the organisation of the Association ('No information'); that Ledoux be delegated to report on the organisation of exchange of material ('No information'); and finally, 'Up to the date of
this letter your secretary has received no word from Dr. LEDOUX, and nothing
pertaining to the Association from Mr. BOULTON.' He does acknowledge
cooperation from Collardet who had kept him informed of the program of the
Congrès International du Bois et de la Sylviculture which was to provide the
framework for the second conference of the wood anatomists.

Record circulated copies of his first draft of the 'Polyglot Glossary of
Terms Used in Describing Woods' to members of the Committee and others
for their corrections and suggestions and expressed the hope that he would be
able to send revised copies to members of the Organizing Committee in time
for the meeting in Paris. A notice of the second meeting of wood anatomists
in Paris was published in *Tropical Woods* to the effect that the Organizing
Committee invites 'all scientists and scientific institutions interested in this
subject to participate in this conference either personally or in writing. ... The
Paris representative of the Committee is Mr. Jean Collardet. ...'

Record did not attend the Paris meeting and Rendle acted as Deputy Secretary in his absence during the sessions which were held between July 1 and 5,
1931. Also attending from among the membership of the Organizing Committee were Boulton, Chalk, Collardet, and Ledoux. Committee members

met on July 2 and agreed upon the form of a constitution for the International Association of Wood Anatomists to be recommended for adoption at an open meeting July 4, 1931. The most significant event of the Paris meeting was, of course, ratification of the Constitution submitted for consideration at the open meeting.

Among the several articles of the Constitution we find the name of the organisation was substantiated officially as the International Association of Wood Anatomists; the object of the Association was to advance the knowledge of wood anatomy in all its aspects; and the activities of the Association were to interchange ideas, facilitate the collection and exchange of wood specimens, work toward standard terminology and descriptions of woods, stimulate the publication of scientific articles, and encourage instruction and study of wood anatomy. There were to be three classes of members: Ordinary Members, Corporate Members, and Honorary Members. The first group comprised those persons actively engaged in the study of wood anatomy; Corporate Members comprised institutions with an interest in wood anatomy; and Honorary Members, persons, who, in the opinion of the Council, had rendered notable service to the advancement of knowledge of wood anatomy. The admission of members was to be controlled by the Council all of whom had to be nominated in writing by two members of the Association. The Council was to consist of no more than twelve Ordinary Members elected by a majority vote of the Ordinary Members. The Council was to appoint a Secretary-Treasurer to be responsible directly to the Council. Council was empowered to fix subscriptions (membership dues) and to have the right to make bylaws for carrying out the terms of the Constitution. Alterations to the Constitution would require a vote of two-thirds of the Ordinary Members of the Association.

Record's next move was to establish a group of Founder Members who would then elect a Council from among themselves. The Council would appoint a Secretary-Treasurer following which activities of the Organizing Committee would cease and Record would be relieved of his duties as Secretary to that organisation. To these ends Record addressed the members of the Organizing Committee on August 13, 1931. He asked them to consider themselves as the first members of the Association and requested them to nominate prominent wood anatomists for membership. He stressed the importance of these first nominees as they 'will be in control of the Association at the most critical stage in its existence. Our responsibility is a serious one and we must limit our first choice to scientists who are outstanding in the field of wood anatomy and about whose qualifications and fitness there can be no question. The members of our Committee are the real founders of the Association and our task is not completed until a Council is elected and functioning. ... I am inclined to believe that all applications for membership should originate with the members themselves rather than with the candidates themselves.'

Record instructed the nine members of his committee to proceed at once to nominate candidates for membership and to send him the full names, addresses, positions, nationalities, and statements of the qualifications of each nominee. Record had an imperious soul. As soon as he had collected information on all nominees he prepared ballots containing all necessary data and circulated them to the committee members for voting. 'Five (5) favorable ballots shall be necessary for election. One (1) unfavorable ballot shall prevent election.' He then notified each person elected and secured his acceptance of membership in writing. After all acceptances from newly elected members had been received, he forwarded to each a 'complete and annotated list of members of the Association' and invited each 'to vote for nine (9) members of the Council.' The nine members receiving the largest number of votes were to be considered elected to the Council. Record notified the first Council of their election and at the same time enclosed a ballot to effect the election of a Secretary-Treasurer.

Polls for election of Ordinary Members were closed December 21, 1931, 36 Charter Members or Founders, representing 15 countries, were declared elected (see *Tropical Woods* 29: 30, 31. 1932 for a list of these Founder Members). Members immediately proceeded to elect the first Council which consisted of 11 councillors owing to a triple tie. These were: M.B. Welch, Australia; Paul Ledoux, Belgium; Laurence Chalk and Bernard J. Rendle, England; Jean Collardet, France; G. Bredemann, Germany; Ryôzô Kanehira, Japan; H.H. Janssonius, The Netherlands; P. Jaccard, Switzerland; and Irving W. Bailey and Samuel J. Record, U.S.A.

Record was elected first Secretary-Treasurer by the Council. Council set subscriptions for Ordinary Members at $1.00 U.S. per annum for the first three years. The Secretary-Treasurer was authorised to collect subscriptions and also to pay expenses incurred in the conduct of official business from the treasury. Two Honorary Members were elected by the Council: Henri Lecomte, Paris, and J.W. Moll, Groningen, The Netherlands. It was agreed at this time that 'All committee reports, standards for terminology and description, manuals, *etc.*, which may be approved by the Council are to be considered merely as recommendations or suggestions for the general guidance of the members.' Even 50 years ago wood anatomists considered themselves an independent lot of free-minded scientists!

To further the stated objectives of the Association, Record continued his work on the 'Polyglot Glossary of Terms Used in Describing Woods.' The first edition which he compiled was given a limited, though international, distribution for review in February, 1931. In a letter to Chalk of March that same year, Record detailed some of his general ideas which relate to the relative value of different sources of information and criticism:

> The next step to try and work out standards for terminology and descriptions and this is best accomplished by getting a representative group of various nationalities to cooperate. We need new terms and new definitions of old ones, but to have each investigator attempting this on his own responsibility will only make more confusion. I invented a number of new terms for the glossary, but I am giving other persons a chance to criticize before putting the new names into circulation.

Record had fixed views of who might contribute significantly to the glossary. He felt there were two groups of workers among the anatomists and indicated that,

> Only the aristocracy of the profession is able to contribute anything toward such a manual. Opinions of those who are not experts are worse than useless. Hence a general association could not be trusted with such a task [*i.e.*, the critical writing of a glossary of terms and definitions]; the more democratic a government is the lower its intelligence! If this manual is to be a scientific contribution it must be prepared by a few open-minded [!?] scientists who know enough at least to understand what is needed and how to go about getting it. The first problem is the task of a small committee, the second for a large and widely representative one. The latter is largely a matter of diplomacy.

Record then gave some examples of the non-critical 'help' he had received from some wood anatomists to whom he had sent copies of his glossary for the sake of public relations.

It is apparent that Samuel Record had a considerably high opinion of himself and of his abilities and a considerably low opinion of some other wood anatomists and their qualities. But, his self-confidence was not entirely ill-placed since in the early 1930s there was little question that he had more experience and knowledge of wood anatomy, expressed through his many articles and three books, than probably any other wood anatomist of the time, save, perhaps, J.W. Moll and H.H. Janssonius.

Record did not see the need for a formal committee to assemble and advise on the preparation of the glossary. Rather, he felt the most expeditious way to assemble the glossary was to 'do the thing informally, getting cooperation wherever it can be had. ... This is exactly what I am doing,' he wrote to Chalk and Rendle in April, 1931. Nevertheless, he agreed, if a formal Association were to be established, then of course, 'we can still have a committee appointed to prepare the manual and this committee can secure all the assistance it needs.'

The informal manner in which Record had been working on the glossary caused some offense, especially since, in effect, he had appointed himself Chairman of the informal committee. There transpired in the correspondence between Record and Chalk and Rendle considerable discussion of whose prerogative it was to appoint Association committees. Record contended that as Secretary-Treasurer it was his right to recommend the appointment of certain *ad hoc* committees to the Council for their consideration and that by doing so he was merely carrying out his official responsibilities and not taking on special privileges. In a letter of June 6, 1932, Record went to great lengths to explain to Chalk why this was so, particularly with reference to the preparation of a glossary of terms. Record wrote:

> Do you realize that all this work on the glossary is still *outside* the Association? ... I had intended making it official business by declaring it so and appointing myself chairman of a committee on terminology. You and Rendle object to my doing it that way. I dislike to go against your wishes and at the same time I dislike to go back to the Council for authority it has already voted to me. The way to avoid both is to continue to keep it [*i.e.*, the glossary committee] unofficial until it is ready to submit to the Council. This would be to the financial advantage of the Association, since none of the cost for typing and postage would come out of its treasury. Do you prefer it that way?

What Record had been doing involved his counseling with a group of U.S. wood anatomists, wholly outside the formal organisation of the Association, to help him with the preparation of a glossary. At the same time, Record solicited recommendations from members of the Association and to that end had sent them copies of glossary drafts for their review and comments. In May, 1932, Record met at Yale with Irving W. Bailey and Ralph H. Wetmore of Harvard, Arthur J. Eames of Cornell, and with George A. Garratt of Yale, to consider the various suggestions received from abroad with particular reference to the choice and definition of English terms. Following the May meeting a revision of the English portion of the glossary of terms was sent to the members of the Association, who had previously collaborated with him, for further review of the terms and definitions. Then, in October, 1932, Record and Garratt went to Harvard and there met once more with I.W. Bailey and R.H. Wetmore, along with Robert H. Woodworth and A.J. Eames, to consider all of the proposed terms and definitions in view of the many suggestions received following the May revision.

After the October meeting at Harvard, Record's committee, now the officially sanctioned Committee on Nomenclature, had transformed its latest draft into a formal report containing 108 terms and definitions and submitted

Fig. 5. Irving Widmer Bailey. – Fig. 6. Ralph H. Wetmore. – Fig. 7. Arthur J. Eames. – Fig. 8. Margaret M. Chattaway. – Fig. 9. Harold E. Dadswell, c. 1940.

it to the Council of the Association for its approval. Nearly all of the terms had been officially approved by the Council by April, 1933, 'but the Secretary deferred their publication until certain queries by individual Councilors could be considered at a meeting of the Committee in October, 1933.'

Originally it had been Record's thought eventually to publish a polyglot dictionary in seven languages in addition to English. But, in the interest of expediency the Committee decided to limit its activities to the selection and definition of English terms leaving translations into other languages to scientists of other nationalities (see 'Multilingual Glossary of Terms Used in Wood Anatomy.' Mitteilungen der schweizerischen Anstalt für das forstliche Versuchswesen 40: 1964). The 'Glossary of Terms Used in Describing Woods' was published in *Tropical Woods* 36: 1-13, 1933 (December). Because certain definitions had not been approved by the Council, those which were approved appeared in boldface type; those which were not yet acted upon by the Council were set in italics.

Publication of the Glossary brought sharp words of criticism from Chalk and from Rendle because it appeared, from their viewpoint, the work was largely of Americans. For example, Chalk wrote to Record in January, 1934, shortly after having received copies of the Glossary:

I cannot help feeling that you have made a mistake in presenting the work of our first international committee so as to make it appear solely the work of Americans. I know that it is somewhere explained that 25 members of the Association collaborated and that the Council approved of the Glossary but the general effect is undoubtedly that the work is solely an American production. You and your American colleagues have doubtless borne the brunt of the work and I do not in the least grudge the full acknowledgement of this; but as you have had international co-operation it seems a great pity not to make the most of it seeing that the whole value of the Association depends on its international character.

There is certainly some truth to Chalk's complaint, for the footnote on the first page of the Glossary listed the members of the Committee on Nomenclature, all of whom were Americans: Eames, Bailey, Wetmore, Woodworth, Garratt, and Record (Chairman). Pages 11 and 12 of the 'Glossary', however, contain an explanation of the germination of the 'Glossary' and the attempts made to have it reviewed by an international body of wood anatomists. It is specifically stated there, 'The Committee's first report was submitted in May, 1932 to the members who had collaborated, about 25 in all. The numerous suggestions received were considered at another meeting in October. ...' (In this respect it is interesting to note that the second edition of the 'Glossary' (*Tropical Woods* 107: 1-36, 1957) was prepared by a committee on which no American was represented: Bruno Huber, Didier Normand, Edgar W.J. Phillips, and Laurence Chalk, Chairman. The introduction states, 'After doing all that was possible by correspondence this committee met in person during the I.U.F.R.O. Congress at Oxford in July, 1956, with Mr.B.J. Rendle as an additional co-opted member, and reached agreement on such difficulties as had remained outstanding.' As far as I know, there were no unfavourable remarks addressed to Chairman Chalk concerning the absence of any Americans on this committee.)

Regardless of the composition of the committee which assembled the 'Glossary' and deliberated over the definitions, that publication of a list of terms and definitions, backed up by the I.A.W.A. Council, even though they were only recommendations and not binding, had a great influence on the stabilisation of wood anatomical terminology. Researchers, anxious for resolution of problems of synonymous terms and for clear unambiguous definitions, seized on the 'Glossary' as a means of achieving consistency and continuity. Paper after paper on wood anatomy research bears a statement to the effect that the terms and descriptions used comply with the recommendations of the Committee on Nomenclature, International Association of Wood Anatomists. A major goal of the Association had been achieved in the short time between its founding in 1931 and the production of the 'Glossary' at the end of 1933. Few scientific organisations have been as successful in accomplishing a principle objective in so short a space of time.

Another of the activities commenced under the aegis of the International Association of Wood Anatomists was an attempt to work toward standard terms of cell size in wood. The problems were outlined succinctly in an article published by Margaret M. Chattaway in *Tropical Woods* 29: 20-28, 1932, entitled, 'Proposed Standards for Numerical Values Used in Describing Woods.' There she outlined the need to define terms of cell size so that they would be universally understood. She asserted the only way to do this was to assign numerical values to the terms, values which had been obtained by proper sampling of mature wood material. In that paper Chattaway proposed classes for different cell and tissue measurements and compared them to similar classes established earlier by other researchers. Considered are numerical distribution of vessels, diameter of vessels, length of vessel segments, numerical distribution of xylem rays, width of rays, height of rays, length of fibres and thickness of fibre walls. She recommended her size classes for consideration by the International Association of Wood Anatomists.

The concept of standardising terms of cell size in wood was not a new idea in 1932; to be sure, Chalk, Rendle, Record and other members of the Association were keenly aware of the need for a soundly based system, especially for its value in wood identification. Accordingly, Record wrote to Chalk in July, 1933 saying, 'Since you have not yet volunteered to accept the chairmanship of a special Committee on the Standardization of Terms of Size of Wood Elements, I should like to draft you for this task.' Chalk responded to accept the committee assignment and noted that he would start on the task as soon as he had finished his study of anomalous woods. A Committee on the Standardizations of Terms of Cell Size in Wood was duly appointed and consisted of Irving W. Bailey, Harvard; S.H. Clarke, Forest Products Research Laboratory, Princes Risborough; Paul Jaccard, Pflanzenphysiologisches Institut, Zürich; Samuel J. Record, Yale; G. van Iterson, Technische Hoogeschool, Delft; and L. Chalk, Imperial Forestry Institute, Oxford (Chairman). Original instructions to the committee were to investigate the confusion existing through the arbitrary use of such descriptive terms as 'large' and 'short' with reference to the size of wood elements, and to suggest a means of standardisation so that the terms, with respect to any cell and any dimension, would always have the same meaning and application. The committee issued an interim report which was read at a meeting of the Association held in conjunction with the Sixth International Botanical Congress, Amsterdam, on September 4, 1935.

An apparent controversy about the value of standard terms became evident to the committee following the circulation of a questionnaire to members of the Association. It was found that some members considered the dimensions of elements to be without diagnostic value and that descriptive terms were sufficient; others took the view that actual measurements were essential to a complete description of wood and held that descriptive terms were superfluous. The Committee concluded that while figures are usually preferable for diagnostic purposes, it is often convenient and desirable to use a simple descriptive term, and it seemed certain to them that both measurements and terms of size would continue to be used for some time. Thus, the committee felt justified in suggesting in its interim report, that standard terms of size should be regarded as supplementary to actual figures.

After original studies by members of the committee (*e.g.*, L. Chalk, The Distribution of the Lengths of Fibres and Vessel Members and the Definition of Terms of Size. Imperial Forestry Institute, Institute Paper 2: 1-12, 1936), the committee submitted to the Council for approval two reports which were published in different issues of *Tropical Woods:* 'Standard Terms of Length of Vessel Members and Wood Fibers' (1937) and 'Standard Terms of Size for Vessel Diameter and Ray Width' (1939). Apparently these were the only reports on the subject of cell size terminology sanctioned by the Council of the International Association of Wood Anatomists. Although the long-term effects of these reports have had considerably less influence on the study of wood anatomy than the 'Glossary' of terms, nevertheless, they represent another milestone in achieving the stated objectives of the Association to work toward standard terminology and descriptions.

In 1938, Samuel Record was permitted to retire as Secretary-Treasurer after having made several attempts to do so. He had served the Association officially for a year as Secretary of the Organizing Committee and for six years as Secretary-Treasurer. (Record died in 1945.) On July 1, 1938, the Council elected Laurence Chalk Secretary-Treasurer. Chalk's Annual Report covers the period from January 1938 to June 30, 1939 and contains a table showing the growth of the Association since its founding in 1931. From an original group of 36 Founder Members, the Association could count 116 members in 1939.

Chalk's second report as Secretary-Treasurer did not appear until July 1, 1946 and it concerns the war years between July, 1939 and June 30, 1946. Although it was hoped the Association could continue to function from England following the outbreak of World War II in 1939, it soon became clear to Chalk that his position as Secretary-Treasurer was untenable. Accordingly, it was decided to appoint Ellwood S. Harrar of Duke University's School of Forestry, Acting Secretary-Treasurer. The Association limped along during the war, as might be expected of any international organisation during those bleak times.

By the end of 1945, a movement was on foot, Chalk wrote, to revive the Association and members were circularised from both the Durham and Oxford offices. The membership had become displaced, new addresses had to be established, surface mail proved unexpectedly slow to some countries (Chalk cites 17 weeks for a reply from Argentina), eight members had died, and there was a general malaise, especially in the overrun countries on the European continent. The Association's main problem had been the election of a new Council, the term of office of the former members having expired.

Chalk remained Secretary-Treasurer through 1946 while arrangements were being made for the election of a new officer to fill this post. Harold E. Dadswell, Division of Forest Products, South Melbourne, Australia, was appointed by the Council beginning January, 1947. At the time of his first annual report, December 31, 1947, the Association listed a membership of 126. It had happily recovered from the dissociations of World War II.

Supplement I

Events chronicled in Tropical Woods

During Record's first years as joint organiser, Secretary of the Organizing Committee, and Secretary-Treasurer of the International Association of Wood Anatomists, the medium for published announcements about the Association was the journal, *Tropical Woods,* which he edited. News usually

appeared toward the ends of issues under the simple title, 'International Association of Wood Anatomists.' From 1930 through 1939, Association news
was a more or less regular feature, sometimes two, or even four, notes ap-
pearing per year. (While Record was editor, *Tropical Woods* was issued quar-
terly, each issue being paged separately and designated as a 'Number.' Numbers are usually cited bibliographically as though they were volumes in the
more commonly employed sense.) Following Record's retirement as Secretary-
Treasurer in 1938, and especially after his death in 1945, fewer and fewer
I.A.W.A. announcements appeared in *Tropical Woods*, the last note being
published in 1957. A reason for part of this was World War II and its interrup-
tion of normal communications. In large measure, however, the increasingly
active *Bulletin of the International Association of Wood Anatomists* (now
IAWA Bulletin New Series), especially in recent years, has assumed the infor-
mational (and scientific) role once played by *Tropical Woods*. Nevertheless,
through 1939, *Tropical Woods* contains a chronology, albeit somewhat
disjointed, of the Association's activities. For the record, these I.A.W.A.
notices and official recommendations of terms, definitions, and standards of
cell size are listed below.

1930. Conference on systematic anatomy of wood. 22: 1,2.
1930. International Association of Wood Anatomists. 24: 1-5.
1931. Congrès International du Bois et de la Sylviculture. 25: 27, 28.
1931. International Association of Wood Anatomists. 26: 18.
1931. International Association of Wood Anatomists. 27: 20-23.
1932. International Association of Wood Anatomists. 29: 29-31.
1932. International Association of Wood Anatomists. 30: 41-43.
1932. Progress of work on international glossary. 31: 29.
1932. International Association of Wood Anatomists. 32: 22.
1933. International Association of Wood Anatomists. 33: 29, 30.
1933. Glossary of terms used in describing woods. 36: 1-13.
1934. International Association of Wood Anatomists. 37: 46.
1936. International Association of Wood Anatomists. 46: 68.
1937. Standard terms of length of vessel members and wood fibers. 51: 21.
1939. Standard terms of size for vessel diameter and ray width. 59: 51, 52.
1945. International Association of Wood Anatomists. 84: 25.
1946. International Association of Wood Anatomists. 86: 25.
1946. International Association of Wood Anatomists. 88: 36, 37.
1957. International Association of Wood Anatomists. 106: 105.

Supplement II

Chronology

1. Informal conference on systematic anatomy of wood held at University of Cambridge, August 18-21, 1930, on the occasion of the Fifth International Botanical Congress, L. Chalk, S.J. Record, B.J. Rendle, conveners. Name established (unofficially) as International Association of Wood Anatomists. Organizing Committee with S.J. Record as Secretary appointed.
2. Second conference on wood anatomy held in Paris during the Congrès International du Bois et de la Sylviculture, July 1-5, 1931. Constitution adopted July 4, 1931; name officially accepted.
3. Thirty-six Charter or Founder Members elected, December 21, 1931.
4. First Council of 11 members elected; triple tie for statutory 12th member broken by vote of elected Council, 1932.
5. Samuel J. Record, Yale University, elected by Council first Secretary-Treasurer, 1932.
6. Draft of 'Polyglot Glossary of Terms Used in Describing Woods' compiled and distributed for review by Record, February, 1931. English version evaluated at Yale, May 27, 28 and at Harvard, October 28, 29 in which terms and definitions were considered in light of suggestions from Association members.
7. Committee on Terminology *(sic!)*, as its first report to Council, submitted a glossary of 108 terms used in anatomical description of woods resulting from October, 1932 meeting.
8. 'Glossary of Terms Used in Describing Woods' published, December 1, 1933; Samuel J. Record, Chairman, Committee on Nomenclature.
9. 'Standard Terms of Length of Vessel Members and Wood Fibers' published 1937.
10. Record retires as Secretary-Treasurer, 1938, after service of one year as Secretary of Organizing Committee and six years as Secretary-Treasurer of the Association.
11. Council elects Laurence Chalk, University of Oxford, as second Secretary-Treasurer, July 1, 1938.
12. 'Standard Terms of Size for Vessel Diameter and Ray Width' published 1939.
13. Secretary-Treasurer shifts to U.S.A. temporarily owing to outbreak of World War II. Ellwood S. Harrar, Duke University, appointed Acting Secretary-Treasurer, 1939.

14. Reorganisation of Association following 1945; Secretary-Treasurer is transferred back to Chalk.
15. Samuel J. Record dies, February 3, 1945.
16. Chalk retires as Secretary-Treasurer, December 1946; Harold E. Dadswell, Division of Forest Products, South Melbourne, elected third Secretary-Treasurer, January, 1947.
17. Association meets with Dadswell at Oxford, July 21, 22, 1947 to consolidate goals following disruption of war.
18. A. Frey-Wyssling, Eidgenössische Technische Hochschule, Zürich, becomes fourth Secretary-Treasurer, 1957.
19. New 'International Glossary of Terms Used in Wood Anatomy' published in 1957; Laurence Chalk, Chairman, Committee on Nomenclature.
20. 'Multilingual Glossary of Terms Used in Wood Anatomy' published in 1964 (English, French, German, Italian, Portuguese, Spanish, Croato-Serbian).
21. Constitution amended, April 6, 1963; September 1, 1970; July 1, 1972; August 1, 1977. Following the 1970 amendation, title of Secretary-Treasurer becomes Executive Secretary.
22. Wilfred Côté, State University of New York, College of Environmental Science and Forestry, Syracuse, elected fifth Executive Secretary, 1970, and Carl de Zeeuw becomes Deputy Executive Secretary.
23. Pieter Baas, Rijksherbarium, Leiden, elected sixth Executive Secretary, 1976, and Peter B. Laming becomes Deputy Executive Secretary.
24. 'Index Xylariorum. Institutional Wood Collections of the World. 2,' published under the aegis of the I.A.W.A., 1978, William Louis Stern, editor and compiler.
25. 'Wood Identification: An Annotated Bibliography' compiled by Mary Gregory, Jodrell Laboratory, Royal Botanic Gardens, Kew, and published by the I.A.W.A., 1980.
26. 'Standard List of Characters suitable for Computerized Hardwood Identification' assembled by an I.A.W.A. committee coordinated by Regis B. Miller, U.S. Forest Products Laboratory, Wisconsin, and Pieter Baas, and published by the I.A.W.A., 1981.
27. Fiftieth Anniversary of the I.A.W.A. celebrated on the occasion of the XIII International Botanical Congress, Sydney, Australia, August, 1981, with symposium on 'Wood Anatomy in Botanical Research,' published as 'New Perspectives in Wood Anatomy.'

Acknowledgements

I would like to thank my daughter, Susan M. Fennell, for her careful sorting and ordering of I.A.W.A. archival materials and preparation of new folders to replace those which had become tattered and friable with age. Doctor C.R. Metcalfe was most kind in offering biographical notes on B.J. Rendle, in reading a draft of the manuscript, and for commenting on I.A.W.A. founding events which he witnessed firsthand. John Brazier, Princes Risborough, similarly helped with material on B.J. Rendle. I owe a great debt of gratitude to Bernard J. Rendle, himself, who most kindly read a draft of this manuscript, as no one else could have, and offered his observations, opinions, and corrections for which I am deeply grateful. Lastly, I want to express my thanks to Dr. Wilfred Côté, for having read the manuscript, for offering useful suggestions, and particularly, for having read the paper during the 50th anniversary symposium of the I.A.W.A., Sydney, Australia.

Systematic, phylogenetic, and ecological wood anatomy — History and perspectives

PIETER BAAS

Rijksherbarium, P.O. Box 9514, 2300 RA Leiden, The Netherlands

Summary: Following an account of the early beginnings of wood anatomy in the seventeenth century, with special emphasis on the often neglected role of Antoni van Leeuwenhoek, history and achievements of systematic, phylogenetic, and ecological wood anatomy are reviewed. Most current concepts in systematic wood anatomy had been formulated towards the end of the nineteenth century. A discussion is given of the major trends in xylem evolution (the 'Baileyan trends'), and of possibilities of reversions of these trends, especially in vessel element (or cambial initial) length. Recently surveyed data on the fossil pollen record of angiosperm families (by Muller) are compared with the occurrence of primitive or advanced wood anatomical characters in extant representatives of these families. Families of Cretaceous origin (as deduced from fossil pollen records) show a much higher incidence of primitive vessel, fibre and parenchyma features, but in families of Tertiary origin the same degree of specialisation occurs as in the extant world flora. For ray type and storied structure no such trend exists. In a discussion of the achievements of ecological wood anatomy, some general trends on vessel element and fibre dimensions, and vessel perforation type as related to temperature and drought are reviewed. Tentative trends for ray histology, parenchyma abundance and fibre type are explored. Rigid adaptionist interpretations are criticised: in addition to adaptive ecological trends, functionless trends imposed by correlative restraints and *'patio ludens'* variation (*sensu* Van Steenis) are advocated. Priorities for future research, in order to promote further integration of the various aspects of comparative wood anatomy, are listed. These are of equal significance in pure and applied research.

Introduction

Systematic, phylogenetic, and ecological wood anatomy collectively are major aspects of comparative plant anatomy and seem to be going through a phase of renewed interest and productivity. In this paper I will attempt to survey the

achievements of these three subdisciplines from their early beginnings onwards, and to indicate possibilities and priorities for future research. The historical parts do not aim at completeness but are meant to give an insight into
opinions of the past based on contemporary knowledge. An admirable history
of wood anatomy in general is provided by Schmucker and Linnemann
(1951), and I have intentionally given emphasis on different aspects in order to
provide a complement to, rather than a repetition of the same story. Much
inspiration can be derived from studying old, only seemingly obsolete, theories and observations. Not only does it prevent unnecessary duplication of
research, but in many cases it proves that old perspectives can be re-adopted
as new.

Early beginnings

If we restrict the study of wood anatomy to microscopic observations, its beginnings are in the seventeenth century with the account of charcoal of several
species and a petrified wood by Robert Hooke in his famous 'Micrographia'
of 1665. Hooke noticed pores of different size classes (vessels and fibres) and
commented that in living wood these are filled with 'the natural and innate
juices of those vegetables.' Whatever the merits of Hooke, the true fathers of
plant anatomy are Marcello Malpighi (1628-1694), Nehemiah Grew (1641-
1712), and Antoni van Leeuwenhoek (1632-1723).

Grew's and Malpighi's work and significance is well known and widely ac-
claimed (Sachs, 1875; Arber, 1941a & b; Metcalfe, 1959 & 1979). Their major
works are also fairly well accessible thanks to the editions of collected plant
anatomical essays in Grew's 'The Anatomy of Plants' of 1682 containing earlier contributions published between 1672 and 1676 and in Malpighi's 'Opera
Omnia' of 1686 containing his Anatomes Plantarum Idea of 1671 and Anatomes Plantarum pars Prima and Altera of 1675 and 1679. Malpighi's botanical writings were moreover partly translated and summarised in German by
Möbius (1901). Both Grew and Malpighi contributed formidably to the understanding of three-dimensional wood structure and illustrated and described a
considerable number of hardwood species and some softwoods. Especially
Grew's work witnesses a comparative and systematic approach; if he describes
wood vessels (thought to be air-filled) he discusses the different sizes and
distribution patterns encountered in different species, and he also observes
that rootwood vessels are generally wider than those of the trunk. In 'An Idea
of a Philosophical History of Plants' of 1672, Grew clearly stated his aims,
and these include the search for common and distinguishing anatomical fea-

tures, which he finds crucial to the understanding of their causality. It is as though Grew gives a precursor to modern speculations on the adaptive significance of wood structural diversity, but probably he meant nothing of the sort. Grew can be credited for having initiated systematic anatomy, because of his awareness of the potential significance of anatomical diversity.

Both Grew and Malpighi were strongly interested in functional aspects, especially concerning conduction. Both considered vessels to function like the tracheae in insects or the bronchieae in mammals for the transport of air; hence their (especially Malpighi's) mistaken view that all woods have vessels with spiral thickenings, mostly in concentric bands at various distances from the pith, in analogy with the annularly reinforced tracheae or bronchieae.

Independent from the above Italian and English Fellows of the Royal Society, Antoni van Leeuwenhoek began to contribute to wood anatomy from 1673 onwards, at the age of 41. Cloth merchant, civil servant and wine-gauger at Delft, in Holland 'the ingenious Mr Lewenhoeck' (Grew, 1682) had also developed great skills in lens grinding, and, with a never ending zeal and curiosity, he not only discovered microbial life and spermatozoa of men and animals, but also made detailed and accurate observations on the structure of wood and bark. Leeuwenhoek's significance for plant anatomy is usually underrated, probably through degrading remarks in for instance Sachs' authoritative history of botany (1875) where it is stated that Leeuwenhoek's work leaves the embarrassing impression of chaos and dilettantism, a view echoed by my fellow country-man Baas Becking (1924) and traceable in Mägdefrau's verdict (1973) that Leeuwenhoek was not led by scientific motives but by sheer curiosity (as though curiosity is not at the basis of all our scientific endeavour). In Leeuwenhoek's lifetime opinion of his work was different. The Royal Society of London elected him a Fellow in 1680, and several other learned societies honoured him. No one less than Robert Hooke complained in 1692 when discoursing on 'the Fate of Microscopes' that they 'are now reduced almost to a single Votary, which is Mr. Leeuwenhoek; besides whom, I hear of none that make any other Use of that Instrument, but for Diversion and Pastime' (cited from Dobell, 1932). Whilst Hooke and Grew did their observations with the compound microscope, Leeuwenhoek used and perfected his simple microscopes – he manufactured about 550 lenses in his lifetime – with magnifying powers of at least $\times$ 170, but probably up to $\times$ 500 (Heniger, 1973) and of great clarity. Comparatively isolated, and initially unaware of the work of his Italian and English contemporaries, Leeuwenhoek pictured and described the anatomy of numerous hardwoods and some softwoods, and was the first to include tropical species (ebony, lignum vitae, cinchona, cinnamon, nutmeg, and coconut) in his research materials. He

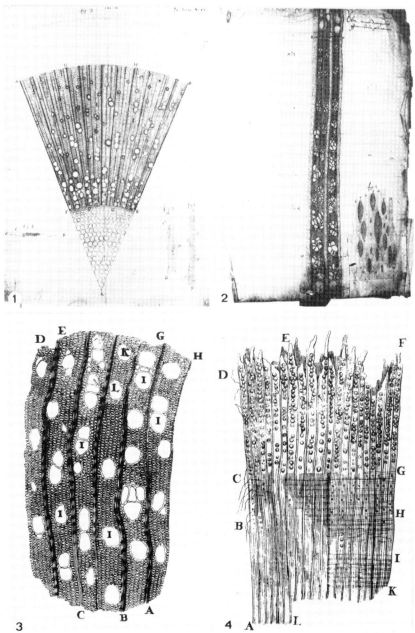

Fig. 1-4. Illustrations by Leeuwenhoek (1 & 2 copied from original letters to the Royal Society; 3 & 4 engravings published in Leeuwenhoek's lifetime by the Royal Society). – 1. Branch of Ash; note detail of vessel with oblique perforation rims at C and D, and numerous bordered pits, termed 'globules' by Leeuwenhoek (from letter of 21 April 1676). – 2. Transverse and tangential

should be credited for a first correct description of bordered pits ('globules' in the walls with a light central dot), perforation rims in vessels, and even a correct observation of the macrofibrillar arrangement in the vessel walls of nutmeg rootwood. Although his ideas on wood ontogeny were totally wrong – he imagined the new wood and bark to be produced from the wood rays, while his contemporaries were just a little nearer the truth by assuming that the wood was formed by the bark – he had a shrewd understanding of tree growth and the influence of climatic conditions on growth ring width and mechanical properties. He realised why ring-porous wood of fast-growing trees was of better quality and higher density than that of slow-growing trees, and why in softwoods the reverse holds true. With Leeuwenhoek we thus have a prelude to ecological and technological wood anatomy. He also appreciated why distinct growth ring boundaries were absent in tropical species. Like Grew and Malpighi, Leeuwenhoek was inspired by the idea of uniformity of structure and function in animals and plants, and in his case this resulted in the mistaken interpretation of all elongate elements as vessels comparable to blood vessels in animals. Thus perforation rims were mistaken for valves, as well as the tangential end walls of procumbent ray cells. Initially he correctly assumed that the true wood vessels serve for upward sap transport, but later he changed his opinion and mistook them for air vessels like his contemporaries and later students of plant physiology (far into the nineteenth century). Tertiary spiral thickenings, longitudinal vessel walls as seen in optical section, the location of intercellular substance between fibres and tracheids, as well as the fibrillar make-up of the cell walls were taken for sap vessels, each with its special sap to fulfill a specific function, such as the nutrition of various parts of seed and fruit. Because Leeuwenhoek's contributions to plant anatomy are generally underrated, and his work is not so easily accessible since he wrote it down in long letters, mainly to the Royal Society, and usually including a variety of topics, I include some of Leeuwenhoek's wood anatomical illustrations (Figs. 1-4) here as a self-evident witness that he deserves to be ranked on the same level with Malpighi and Grew as one of the three founders of plant anatomy. Leeuwenhoek's letters are being re-edited this century (Leeuwenhoek, Collected Letters, 1939-1982). A more or less comprehensive survey of his plant anatomical contributions is given elsewhere, with full references to subsidiary

section of Elm; note two growth ring boundaries and uniform vessel distribution pattern in transverse section (from letter of 12 January 1680). – 3. Transverse section of Nutmeg *(Myristica)* rootwood; note realistic portrayal of angular vessels (from letter of 1 May 1695). – 4. Radial section of Spruce with torn ends; note small crossfield pits and numerous bordered pits in tracheids (from letter of 12 August 1692). – All figures from Leeuwenhoek's collected letters (1939-1982), Swets & Zeitlinger, Amsterdam.

literature (Baas, 1982). Biographical data can be found in Dobell (1932), Schierbeek (1950, 1951, 1959), and Heniger (1973).

Whatever the mistaken functional interpretations of Malpighi, Grew, and Leeuwenhoek, their observations on wood structure were to remain unrivalled for about a century. Kieser (1815) ably reviewed the early plant anatomical scene when he stated 'Hooke gibt nur einzelne aber brauchbare microscopische Abbildungen; Grew ist am elegantesten; Malpighi am ausführlichsten; Leeuwenhoek am treuesten. Malpighi und Grew haben sich oft von vorgefassten Meinungen hinreissen lassen; ihre Werke sind systematisch. Leeuwenhoek gibt nur isolirte Untersuchungen, aber reiche, oft verkannte Beiträge zur höhern Pflanzenanatomie.'

The eighteenth century would witness little progress in wood anatomy. Duhamel du Monceau (1758) provided no anatomical novelties, but published decisive experiments on the ontogeny of wood and bark in his 'Physique des Arbres' (fide Boerlage, 1875, and Schmucker and Linnemann, 1951). John Hill (1770) in his treatise on the construction of timbers also added mistakes rather than understanding, but he should be credited for first isolating fibres by maceration in water. However, this led him to the most negative remark one can think of for the field of comparative wood anatomy: 'tho' difficult to obtain, there is little for observation in it when we have it; only that truth is always valuable; and when we know the composition of the wood in one tree, we can understand in all.' Yet he foreshadowed systematic wood anatomy when commenting on similarities and differences in the vessel distribution of various oak species. Van Marum (1773) largely followed Leeuwenhoek's interpretation of wood elements in his thesis on liquid transport in plants (fide Boerlage, 1875).

Following the long dormant period, plant anatomy was to experience a revival from the onset of the nineteenth century onwards. The first half of the century was needed to settle the controversies and solve the problems inherited from Grew, Malpighi, and Leeuwenhoek. Sachs (1875) and Boerlage (1875) have simultaneously given comprehensive accounts of the developments of this period. They clearly showed that despite the great significance of men like Bernardi, Kieser, Mirbel, Sprengel, Link, Treviranus, Moldenhawer, Meyen, etc., progress was by no means a steady advance but involved many retrograde steps if compared with earlier and better interpretations of their seventeenth century predecessors. The greatest achievements for our discipline were obtained by Hugo von Mohl (1805-1872). This excellent microscopist not only made a point of executing his own drawings (until then customarily left to draughtsmen), but also developed – like Leeuwenhoek – a great skill in grinding his own microscope lenses (fide Sachs, 1875). Gradually agreement on the nature of vessels and the structure of bordered pits was obtained, also thanks to developmental studies. Such an agreement was a prerequisite for any large-scale application of the knowledge of wood structure in a comparative sense. Once obtained, systematic wood anatomy could enjoy its explosive development in the second half of the nineteenth century.

Systematic wood anatomy

The potential use of wood anatomical diversity for taxonomic purposes had already been hinted at by various authors who had primarily concerned themselves with the elucidation of wood structure for its own sake in the first decades of the nineteenth century. Kieser (1815) for instance, had concluded his 'Phytotomie' with the remark that 'das grösste Desiderium, dessen Erfüllung einziger Zweck der Pflanzenanatomie sein muss, ist also eine vergleichende Anatomie; und ehe diese gegeben ist, kann keine genügende Classification der Pflanzen ... vollendet werden.' De Candolle (1818) had used presence or absence of vessels in his plant classification. Yet, critical studies based on extensive research materials had to wait until the second half of the century.

Hartig (1859) grouped numerous genera on various combinations in which wood elements (rather badly classified by him) can occur, and had given as his opinion that wood anatomy could eventually lead to a replacement of classification systems based on traditional features ('Äusserlichkeiten', *i.e.*, superficial external characters). Sanio (1863), who studied about 150 genera, arrived at the more balanced opinion that wood anatomical features have a relative systematic value, like all other morphological characters. He based this opinion on patterns of variation he observed in obviously related taxa, such as species of one genus and genera of one family, and noted that characters constant in one assemblage could vary in others. Similar conclusions were arrived at by Boerlage (1875) who studied the tribe Artocarpeae of the Moraceae. The overoptimistic views of Hartig were shared by Chalon (1868), who claimed that all natural plant groups have their own wood structure, constant for all species, and usually sufficiently diagnostic to recognise genera. Such statements were contradicted by pointing out similarities between wood of the temperate representatives of Leguminosae and *e.g.* the unrelated *Morus alba* (Boerlage, 1875). Möller (1876) took a very critical view of the systematic value of some wood anatomical characters and claimed, for instance, that tracheids were mere vessels without perforation, and that the lack of spiral thickenings was simply a consequence of their diameter and thus had no taxonomic value (*cf.* Van der Graaff and Baas, 1974, for reminiscent views). Vesque (1881) also maintained that wood characters, especially those of the vessels, were unreliable features for taxonomic purpose, and could only be used when comparing plants of similar environments.

Truly large-scale wood anatomical comparisons were stimulated by Radlkofer, who extensively used wood ànd leaf anatomical characters in his taxonomic studies on the Sapindaceae. His student Hans Solereder was to take a PhD degree on a thesis entitled 'Über den systematischen Wert der Holzstructur bei den Dicotyledonen' in 1885. In this thesis constancy and variability of all major wood features were tested on a sample of 1200 species belonging to 140 families. Solereder had to content himself with herbarium twigs, which makes it understandable that his emphasis is on vessel and fibre characters (especially types of perforation plates and pitting) and not so much on parenchyma distribution and ray composition. Solereder's thesis ought to be compulsory literature for any modern student of comparative wood anatomy. Not only are his conclusions on the relative, but often considerable systematic value of various wood anatomical features still perfectly valid, but he also gives such a wealth of wood structural information that he can stand comparison with our present stage of knowledge on diversity in wood anatomy with glamour. For instance, on vessel perforation anomalies (crossed bars, combination simple-scalariform plates, *etc.*) which we only presume to have discovered in our present scanning electron microscopy era, he gave good descriptions and numerous examples. In passing it might perhaps be stressed that in the pursuit of understanding little known structures with modern equipment, one should always turn to the early literature for inspiration. It will appear that structure and ontogeny of most features such as tyloses (Zimmermann, 1979), compression wood (Timell, 1980), and trabeculae (Werker and Baas, 1981) were at least as well understood as we can pretend to do now after sophisticated ultrastructural research.

French and German anatomists were exceedingly productive in the late 1800s. De Bary (1877), in the preface to his excellent textbook on comparative plant anatomy, even complained that he could not possibly include full references to the literature in his text, because then almost each word had to be accompanied by one or several authors' names. That many of these papers were specifically devoted to systematic anatomy can be judged from the references cited in Solereder's 'Systematische Anatomie der Dicotyledonen' of 1899. The publication of this standard work and its complement of 1908 can be considered the culmination of an era, appropriately dedicated by the author to L. Radlkofer and starting with the self confident citation from the latter's Festrede of 1883: 'Die nächsten hundert Jahre gehören der anatomischen Methode.' In his introduction Solereder stressed limitations as well as perspectives of the anatomical method, and pointed to the need of establishing without prejudice the constant occurrence (*i.e.*, the diagnostic value) of anatomical characters for each species, genus and family. He distinguished between two categories of anatomical features: phyletic characters on the one hand, and physiological and biological ones on the other. The former are independent of ecological conditions, and were acquired in unknown times through unknown causes, and have the greater taxonomic value. Physiological and biological characters were acquired as adaptions to climate and habitat, or to animals, mechanical requirements and special demands imposed by the growth habit. The taxonomic value of these characters is mostly limited to the species level. Adherents of the anatomical method should not expect to overthrow the traditional classification systems based on macromorphological characters; mostly anatomy confirms the older classification, but as a very important auxiliary discipline, systematic anatomy can contribute to, or improve systems based on floral and other external characters. Identification of incomplete plant material with characters from vegetative anatomy is an important application of the anatomical method. I have quoted Solereder's views at some length in order to demonstrate that in the late nineteenth century concepts in systematic anatomy were fully mature, and that twentieth century authors have not added significant new principles as far as phenetic systematics using anatomical features are concerned. Because of Solereder's practical restriction to twig material for his wood anatomical descriptions, systematic wood anatomy had to be further developed using samples of mature tree stems. Chalk and Metcalfe (1979) have described and emphasised the roles of institutional wood collections and the significant contributions from institutes of applied wood science in this field. The data accumulated in the first half of our century, with numerous original observations and taxonomic comments, are ably summarised in Metcalfe and Chalk's 'Anatomy of the Dicotyledons' (1950). This most frequently quoted source of reference in the botanical literature is a worthy successor of Boodle and Fritsch' English translation of Solereder's 'Systematische Anatomie' (elaborated by D.H. Scott).

Returning to the nineteenth century, one can trace a noticeable influence of systematic anatomy, including that of the secondary xylem, on classification systems. Engler and Prantl's 'Die natürlichen Pflanzenfamilien', published between 1889 and 1897 in its first edition, includes anatomical details and considerations, and several taxonomists authoring specific family treatments made their own original observations on wood structure (notably Solereder, but also Engler himself in his account of *e.g.* the Saxifragaceae). The great botanists Vesque and Van Tieghem in France also routinely included microscopical features in their taxonomic endeavours, although the former continued to mistrust wood anatomical characters (1889) and preferred various leaf anatomical features such as stomatal type and petiole and midrib vascularisation.

Developments of comparative wood anatomy in the twentieth century have been described in detail in a number of fairly recent papers (Jane, 1963; Brazier, 1975; and Chalk and Metcalfe, 1979). In summary one can signal major achievements along three different, complementary lines: 1. wood anatomical surveys of the major tree (and shrub) species from a specific geographical area – quite often colonies of Western powers in the tropics (*e.g.* Moll and Janssonius' *magnum opus* on Javanese woods of 1906-1938); 2. wood anatomical monographs of individual families (numerous examples included in Gregory's identification bibliography of 1980 under the family headings); 3. analyses of the distribution of specific characters in woody plants and an assessment of their taxonomic value (*e.g.* Bailey's study of vestured pits of 1933, Chattaway's papers on crystals of 1955 and 1956, and numerous contributions by Record in the journal Tropical Woods).

Activities are no longer confined to the European continent and Great Britain but have spread all over the world with fluctuating activities in each centre, depending on the perseverance and drive of individual wood anatomists. Major incentives for systematic wood anatomy came from applied wood science where a great demand for identification of timber samples and predictive knowledge on properties was and is continuously felt. The International Association of Wood Anatomists, founded in 1931, came to play a stimulating role in these developments (Stern, 1982 - this volume). The recent completion of a Standard List of Characters suitable for Computerized Hardwood Identification by an IAWA Committee in 1981 provides an example of how wood anatomical diversity can be classified to be applied to the practical need for sophisticated identification methods in numerous fields of interest. In taxonomy the significance of wood anatomical characters is now widely recognised, and evidence from wood anatomy has pervaded all modern systems of angiosperm classification, like the ones by Takhtajan (1980), Thorne (1976), Cronquist (1968), and Dahlgren (1980). To a great extent this has also been stimulated by the phylogenetic concepts introduced into wood anatomy from 1918 onwards by Bailey and his students (see below).

Meanwhile the debate on the taxonomic value of wood anatomy in general or certain specific features continues along much the same lines as it was carried out in the second half of the nineteenth century. The questions: 'which is the taxonomic and diagnostic value of a certain character', and, 'at which level (species, genus, family, order) can each wood anatomical variable be used', have to be answered for each individual taxonomic group. If at all there is a generally applicable principle in systematic wood anatomy, it is the rather negative one that no absolute generalisations can be made on the value of any single character. In this respect the wood anatomical approach does not differ from any other source of enquiry in plant taxonomy. A character like vestured pits can help define entire orders or large families as Myrtales, Rubiaceae, and Leguminosae, but may be variable in closely knit groups even below the genus level in Boraginaceae (Miller, 1977) and Cistaceae (Baas and Werker, 1981). In the latter category of cases the gradual variation of conspicuous presence, via rudimentary (or incipient?) presence, to total absence opens interesting possibilities to hypothesise on the phylesis of such characters. Similar examples could be given for each character, notably those involving vessel perforations, arrangement of lateral wall pitting, fibre type, parenchyma distribution and ray type.

The evaluation of taxonomic and diagnostic value can be made by the simple test of constant occurrence in the taxa of different rank, without consideration of ecological or phylogenetic aspects. However, research materials are usually restrictive here, because for many woody species wood samples are scarce or completely wanting in institutional wood collections, and their collecting in the field may meet with obstacles or intolerable delays for a research project of restricted duration. Systematic wood anatomists therefore have to make use of insufficiently tested discontinuities in the observed variation patterns, as the lesser of two evils (the other evil being: not to use any wood anatomical evidence at all).

There are two fundamentally different approaches to exhaust the potential of wood structural variation for plant taxonomy.

In the first approach those features are eliminated from the taxonomic discussion which are likely to be strongly influenced by within-a-tree or within-stand variation, or depend too much on abiotic factors like climate and soil. Most quantitative features are known to depend to various extents on these factors, and several qualitative features such as degree of ray heterogeneity may gradually vary throughout the entire life-span of a tree (*e.g.* De Bruyne, 1952). The remaining characters such as vessel grouping, vessel perforation type, vessel and fibre wall pitting, main type of parenchyma distribution and rays, inclusions, secretory elements, and large discontinuities in quantitative features resulting in almost qualitative differences in overall xylem histology, have been sufficiently tested in numerous plant groups, and are likely to be of taxonomic significance in the plant group under study. Within these characters one may even indicate some hierarchical order, and vessel perforation type and fibre type can be given more weight than the other characters. Although this procedure may seem in conflict with my earlier remark that no *absolute* general rules apply in systematic wood anatomy, it is usually a fruitful approach resulting in classifications improving or even matching those based on *a combination* of features from other plant parts. It has been the tacitly adopted practice in systematic wood anatomy from its nineteenth century beginnings onwards, and will retain its value as long as unbiased (*i.e.*, verifiable) arguments are forwarded why each character has been given its assigned weight.

In the second approach all varying features recorded in the descriptive phase are given equal weight, and the classification is based on computer-aided pattern detection analyses of various sorts. In plant taxonomy the possibilities of numerical analysis have been widely explored; in wood anatomy there are only few studies to date applying this approach. Koek-Noorman and Hogeweg (1974) were to my knowledge the first to apply cluster analysis, using 125 wood anatomical characters in three tribes of the Rubiaceae. The most interesting result of their studies is that irrespective of whether all characters are given equal weight, or whether preferential character weighing is applied, the wood anatomical classification remains virtually the same. Only one set of characters (*i.e.*, those of the fibre pitting) appeared to be true to the major bipartition of their classification. Deletion of this 'heavy character' from the cluster analysis does, surprisingly enough, not appreciably alter the dendrogram. These results imply that, although many characters included in the data base would be considered to introduce a high noise level (*i.e.*, represent variable characters of restricted taxonomic significance), all characters contribute to a consistent wood anatomical classification. In a later paper Hogeweg and Koek-Noorman (1975) studied the effects of iterative character weighing (*i.e.*, weights obtained iteratively on the basis of the distribution of character states in previously generated classes) and found basically a similar classification for the tribes of Rubiaceae concerned. Reassuringly enough the character weights obtained by this autonomous (*i.e.*, non-supervised) method did not run counter to intuition. The success of the numerical methods obtained for the Rubiaceae and later for the Melastomataceae (Koek-Noorman *et al.,* 1979) is probably due to the very sensible choice of distinctly different states for each character, implying large size or frequency classes for quantitative characters, thus restricting the noise level in the data base. In my opinion this noise level is too high in *e.g.* the principal component analysis based on nine characters, of which six are quantitative in Robbertse *et al.*'s study (1980) of *Acacia* woods.

With iterative character weighing, computerised numerical pattern analysis has become principally similar to the first approach of traditional systematic wood anatomy, albeit that the computer can handle large data sets more quickly and weigh characters more objectively than individuals can. Yet it is reassuring that in all the studies mentioned, the traditional approach would have led to basically similar conclusions on wood anatomical grouping.

Obviously, wood anatomical variation patterns need to be integrated with data from all other sources of enquiry suitable for classification purposes. Ideally the taxonomist should incorporate the evidence from all subdisciplines in his final classification. In practice, this approach is still rather rare. The recent treatment of Loganiaceae (Leeuwenberg *et al.*, 1980) in Engler and Prantl's 'Die natürlichen Pflanzenfamilien' is a fortunate exception, although even here one wonders whether pollen morphological and wood anatomical diversity have been given enough weight. For wood anatomists it may thus become necessary to almost overstate their case and formally suggest alternative classifications, in the hope that their results will receive sufficient attention from later monographers. Simultaneous team-work by anatomists, pollen morphologists, cytologists, phytochemists, *etc.*, and a taxonomist is of course a much better approach, but can often not be achieved or synchronised.

Phylogenetic wood anatomy

The breakthrough in phylogenetic wood anatomy by Bailey and Tupper (1918) and the further development of the 'Baileyan concepts' have been summarised and discussed *in extenso* several times (*e.g.*, Metcalfe and Chalk, 1950; Carlquist, 1961, 1975; Stern, 1978). Before 1918 wood anatomical features like ray width and ray aggregation had also entered debates on the phylogenetic derivation of herbs from woody plants (see summary in Metcalfe and Chalk, 1950: xxxii-xxxix).

The outcome of Bailey and Tupper's reconnaissance of secondary xylems in vascular Cryptograms, Gymnosperms, and Angiosperms are the well-known major trends of wood anatomical specialisation from a vesselless condition with long tracheids, via wood with relatively long scalariformly perforated and pitted vessel elements and fibre-tracheids, towards wood with short, simply perforated, alternately pitted vessel members and libriform fibres. In later years these trends have been supported by many studies, and could be amplified with general trends in parenchyma distribution and ray histology. It may be interesting to note that Bailey and Tupper's result is far removed from their initial aim, expressed in the introduction of their classic paper: *i.e.*, to contribute to a discussion on the significance of cell size, a topic which intrigued botanists and zoologists alike around the turn of the century. Cell size, or rather fusiform element length, is the corner stone of Bailey and Tupper's elegant series; yet it is gratifying that also without this quantitative character involved, the morphological series from tracheid to specialised vessel member makes sense and finds support in the fossil record (see below).

The impact of Baileyan trends on the phylogenetic classification of Angiosperms has been considerable, especially in the choice of Magnoliales (or woody Ranales) instead of the Amentiferae as extant Angiosperms which have retained the most primitive features. Dickison (1975) has concisely reviewed the general agreement between phylogenetically interpreted wood anatomy and the classification systems of Cronquist and Takhtajan.

The irreversibility of the major trends as claimed by Bailey on various occasions, has been challenged to some extent in recent years (Baas, 1973, 1976; Van der Graaff and Baas, 1974; Carlquist, 1975, 1980; Van den Oever *et al.*, 1981). On the basis of ecological trends, and on the assumption that these cannot always have run parallel with evolutionary sequences, reversibility of especially quantitative characters like element length, diameter and frequency has been advocated. It is a matter of opinion whether one wishes to emphasise these possibilities for reversion as I did, or whether it is considered to be highly improbable as a phenomenon affecting more than one character at the same time (Carlquist, 1980). On the whole, I agree with Carlquist's defence of the largely irreversible nature of the major trends as an integrated syndrome of xylem specialisation, allowing for a certain noise level and admitting ecological trends which can even be considered to be the driving force in the otherwise unexplained, autonomous trends. However, there are several reasons (of varying importance) why one should be alert for possible reversions:

1. At least one mechanism, likely to have played a significant part in form making in evolution, *viz.*, hybridisation resulting in allopolyploidy, can produce longer vessel elements in the offspring than are present in the parent species, because ploidy level is related to element length (Armstrong and Funk, 1980).

2. Element length is subjected in most cases to strong ontogenetic increase (the so-called Sanio curve for length-on-age variation). This phenomenon is paradoxical, if one considers that first-formed secondary xylem is usually a refugium of primitive qualitative characters. The proportional increase of tracheid or vessel element length varies considerably: from 1.5-6.4 in Gymnosperms, and from 1.0-3.0 in Angiosperms as calculated from Bailey and Tupper's data on about 120 woods of which both the first-formed secondary xylem and older wood was analysed. It also varies within genera or even species (Baas, 1973, 1976). This variation has probably nothing to do with phylogenetic specialisation, and may be highly reversible. Vessel element length if based on mature samples thus presumably includes an important source of reversible variation.

3. Quantitative features such as element diameter, length and frequency are strongly influenced by ecological trends (acting either on the phenotypic or genotypic component of the variation), which cannot always have run parallel with evolutionary sequences, so that these characters contain too high a noise level and cannot be used for unidirectional phylogenetic trends. Sampling differences may also account for huge differences in these characters, which have nothing to do with phylogenetic specialisation (see also Van den Oever *et al.*, 1981). Where the ecological trends run counter to evolutionary events, reversion is implied as for instance in Hawaiian *Euphorbia* as hypothesised by Carlquist (1970).

4. Paedomorphosis (or juvenilism), an explanation first introduced into wood anatomy by Carlquist (1962) for a special category of woody plants which probably evolved from a herbaceous ancestry, and implying that characters typical of metaxylem and first-formed secondary xylem are perpetuated in the secondary xylem, offers a host of possibilities for reversion of the Baileyan trends. This is because primary and first-formed secondary xylem is usually a refugium for primitive characters such as scalariform wall pitting and often also scalariform perforations and a high proportion of erect ray cells. Carlquist (1962, 1975, 1980) has reserved this mechanism for presumed secondarily woody plants such as giant groundsels *(Senecio), Lobelias* and other pachycauls. Although challenged by Mabberley (1974), Carlquist's paedomorphosis explanation has gained credibility because of its predictive value: several odd woody taxa belonging to largely herbaceous groups, and which on independent grounds one suspects to be derived from a herbaceous stock, appear to show the paedomorphosis syndrome. Once accepted for this category of woody plants, I can see no reason why paedomorphosis cannot also have played a part in diversification of wood structure of primarily woody groups. Admittedly, it remains a matter of speculation, as to what extent paedomorphosis can account for the occurrence of primitive features in otherwise fairly specialised groups, but in some cases it seems at least as good an explanation as the often invoked phenomenon of heterobathmy. Melastomataceae offer an example where rays of largely erect cells in many shrubs and trees have probably nothing to do with secondary woodiness but with paedomorphosis (Van Vliet, 1981). In this family there is another interesting case of possible reversion from the major trends: Van Vliet (1981) provided evidence that in certain Melastomataceae scalariform inter-vessel pits are probably phylogenetically derived from alternate pits.

5. One reason to almost overemphasise the possibilities for reversion is the

38

imminent danger that individual small differences which can be interpreted as small parts of the Baileyan trends are used to demonstrate that one taxon is more primitive than another in the wood anatomical character concerned. I have made this mistake when using data from twig anatomy in *Platanus* (Baas, 1969) to stress the primitive status of *Platanus kerrii*.

With the possibility that in the future cladistic methods will become commonly employed in systematic wood anatomy (*cf.* Baas and Zweypfenning, 1979; Koek-Noorman, 1980), the temptation to use small differences in wood anatomy in order to dispose of more characters for which one can claim a primitive (plesiomorphic) or derived (apomorphic) status will be great, but will almost certainly lead to false conclusions.

6. Some of the arguments in favour of irreversibility, *viz.*, the inability to imagine that perforation plates have increased in bar number, or that there can even be a reversion from simple to multiple perforation plates, and the emphasis on the integrated way in which specialised wood anatomical syndromes evolved, are not very convincing. We do not know the way in which these wood anatomical features are controlled, but if the phylogenetic trends are the result of random mutations or more drastic changes in the genome acted upon by selective pressure, I do not see why genetic changes do not also offer possibilities for reversion in conditions where the more primitive character state is not adverse to optimal functioning or even more suitable than the specialised state. Carlquist (1975, 1980) has hypothesised such functional superiority of primitive features for a number of cases. As for the integrated way in which wood specialisation has occurred: this should not be overestimated. All trends for vessel perforation type, wall pitting, fibre type, parenchyma distribution and ray type have been convincingly established by statistical correlation. However, in all correlations there is still a considerable percentage of non-correlation: in other words in numerous cases (ranging to over 40%) primitive characters are associated with specialised ones (see tables in introductory chapters of Metcalfe and Chalk, 1950). This phenomenon of heterobathmy does not imply reversibility, but weakens the case for irreversibility.

The above, largely theoretical considerations are an extension, not a criticism of Carlquist's statement (1980) that 'The 'irreversibility' of the major trends of xylem evolution as stated by Bailey and others are probably misleading if read in an excessively narrow sense, and therefore are capable of being misunderstood.' In retrospect it is a great tribute to Bailey and Tupper's conclusions of over 60 years ago, that all of their major trends in vessel and tracheid specialisation have been strengthened on subsequent research. Evidence from the fossil record now also unambiguously supports the trends, primarily established on the bases of extant plants.

In Bailey and Tupper's original study knowledge of the secondary xylem of vascular Cryptogams and older Gymnosperms had provided a convincing starting point of the morphological series so that it could be read only in one direction. Chalk (1937) showed that in 49 form genera of fossil woods (Cretaceous ànd Tertiary) there is a much higher incidence of primitive features than in a random sample of about 1260 extant woody genera. A comprehensive analysis of more fossil woods, separated according to geological age, would provide a most welcome addition to test all Baileyan trends (including the more questionable ones on ray and parenchyma specialisation). In this respect the very detailed study of about 70 angiosperm woods from the Upper Cretaceous of Central California (Page, 1979, 1980; Wolfe *et al.*, 1975) is most significant. These woods include two vesselless taxa, and features like scalariform perforations and medium or long vessel elements abound. None of the samples have complex pore arrangements, storied elements, resin canals or tile cells, and axial parenchyma is never in multiseriate or aliform sheaths (Page, 1980). Wood anatomical evidence from the fossil record cannot compete with the more complete data on fossil pollen. Muller (1981) has critically compiled all the information on fossil angiosperm pollen which can unambiguously be identified with pollen types of extant families. On the (questionable) assumption that families with recognisable pollen types have retained more or less the same wood anatomy throughout their evolution I have compared Muller's data with the incidence of certain wood anatomical features in these families and in the woody Dicotyledons as a whole. The latter data are derived from Metcalfe and Chalk (1950) from their analyses of the wood anatomy of 1800 woods (species; but mostly with only one or few species per genus) and from their descriptions of families with woody representatives. Although the latter are in need of updating, inaccuracies and incompleteness cannot have influenced the results of Table 1 significantly. In the table, characters to which great importance has been attached in phylogenetic wood anatomy like vessel perforations, fibre type, parenchyma distribution, ray type and storied structure as well as the phylogenetically 'unimportant' or 'ambiguous' characters, like presence of spiral thickenings in vessels and vestured pits, are listed as percentages of families with their first pollen records in: a. Cretaceous (Aptian to Maestrichtian, 110-69 million years old); b. Paleocene (65-55 million years); c. Tertiary as a whole (65-5 million years) and in the extant woody flora. Percentages are also presented for all families known from the Cretaceous and

Table 1 – Incidence (in percentages) of various wood anatomical features in extant families of different geological age according to their fossil pollen record (Muller, 1981), compared with character frequency in the extant woody flora of the world by families and genera (the latter data from Metcalfe and Chalk, 1950).

	Families of Cretaceous origin (24)	Families of Paleocene origin (15)	Families of Tertiary origin as a whole (c. 83)	Families of Cretaceous and Tertiary origin together (c. 107)	Extant families (195)	Extant genera/species (sample of 1800)
VESSELS						
Vesselless	4	0	0	1	2	0.5
Perforations exclusively scalariform	33	12	7	12	16	11
Perforations scalariform and simple	21	19	21	22	18	5
Perforations very rarely scalariform	17	31	24	22	18 }	-
Perforations exclusively simple	25	38	48	43	46 } 83.5	-
FIBRES						
Typically fibre-tracheids	54	38	26	32	31	33
Intermediate or both fibre-tracheids and libriform	18	19	17	17	14 }	-
Typically libriform	28	43	57	51	55 } 67	-
PARENCHYMA						
Mostly apotracheal	54	27	20	27	34	36
Paratracheal ànd apotracheal	17	40	40	35	29	20
Mostly paratracheal	29	33	40	38	37	44
RAYS						
Heterogeneous	62	47	55	56	67 }	-
Heterogeneous and/to homogeneous	17	33	35	31	21 } 79	-
Homogeneous	21	14	10	13	12	21
STORIED STRUCTURE						
Present	28	20	28	28	28	14
SPIRAL THICKENINGS						
Present in vessels	60	69	57	58	52	9
VESTURED PITS						
Present	13	19	15	15	12	?

Tertiary collectively. Especially for percentages based on a low number of families (*i.e.*, those that had their first pollen record in the Cretaceous or Paleocene) the values are subject to much chance fluctuation. Yet the pattern for vessel perforations, fibre type and parenchyma distribution are consistent in showing a much higher incidence of primitive character states (scalariform perforations, fibre-tracheids, apotracheal parenchyma) in the families of Cretaceous age as compared with the extant flora. For the small number of families that had their (pollen morphological) origin in the Paleocene the pattern is more or less similar to that in the extant flora, with only a slightly higher incidence of primitive characters. However, one is struck by the very close similarity of character distribution for all families of Tertiary origin and the extant flora; this similarity becomes even stronger if the families of Cretaceous + Tertiary origin are taken together — in other words: in all families represented in the fossil pollen record towards the end of the Tertiary the wood anatomical specialisation for vessel perforation, fibre and parenchyma characters is the same as that in the extant woody flora. For ray composition the story is different; the presumably specialised type (homogeneous; *i.e.*, composed of (almost) exclusively procumbent cells) appears to be more common in the Cretaceous families than in families from the Tertiary or the present flora. Here it should be stressed that the difference in percentage is quite low; and the high percentage for Cretaceous families is due to only 5 out of 24 showing homogeneous rays in (part of) their species. For storied structure, a feature associated with a host of highly specialised anatomical features, the same unexpected result applies: families first recorded in the Cretaceous show the same proportion as Tertiary and extant families with this specialised character. For the characters spiral thickenings in vessels and vestured pits there are no appreciable differences in the various categories. The slightly higher percentage for spiral thickenings in families with first pollen records of Cretaceous and Paleocene age can moreover be accounted for by the present wide distribution of most of these families, including temperate regions where many taxa have developed this feature (*cf.* Baas, 1973).

As stated above it is questionable whether one may assume that families recognised by their pollen types have retained the same specialisation level in their wood throughout their evolutionary history; in view of the totally different biological significance of pollen morphology and wood anatomy in ecological diversification this assumption is even highly unlikely although there is a certain degree of correlation of primitive wood structure and primitive pollen morphology in extant plants where for instance all vesselless Angiosperms have monosulcate-derived or unspecialised tricolpate pollen grains. Thus the percentages for families in Table 1 should not be taken as a prediction of the actual distribution of wood anatomical characters in the Cretaceous and Tertiary eras. If the assumption of general irreversibility holds true, the percentages of families with primitive characters would, however, represent minimum values, likely to have been higher at the time of angiosperm diversification in Cretaceous and Paleocene. This finds support in the data on Upper Cretaceous woods from Central California (Page, 1979, 1980) where exclusively scalariform plates occur in over 50% of the entities; the majority has apotracheally diffuse parenchyma, and the rays are almost always hetero-geneous. Exact percentages have not yet been published, but are likely to ex-ceed those for the families of Cretaceous origin of Table 1. Similarly the total absence of storied structure from the Upper Cretaceous fossil woods contrasts with the relatively high percentage in the extant families of Cretaceous origin, and would imply that in these ancient families this character has evolved later in their evolutionary history (almost certainly always in parallel development).

From the inferences based on data of the fossil pollen record, it would seem that towards half-way the Tertiary (Oligocene) wood anatomical specialisa-tion had fully reached its present level. There are, however two additional fac-tors which can explain why in Table 1 the percentages for primitive character states are lower than expected in the Cretaceous and early Tertiary flora. Firstly, the recognisable fossil pollen types constitute only part of the entire fossil pollen record and favours specialised types, because these are the only ones with diagnostic features. A proportion of the fossil pollen record, and to a smaller degree of the extant flora shows unspecialised types with a common occurrence in numerous unrelated families. These unspecialised pollen types tend to be associated with lower wood anatomical specialisation levels (Muller, personal communication), so that the percentages of primitive fea-tures in fossil taxa in Table 1 should be higher. In passing, it may be noted that the pollen morphological specialisation level of the 24 families of Creta-ceous origin is not significantly lower than that of the extant flora (Muller, personal communication); for these families we can thus conclude that their wood anatomy is more conservative than their pollen morphology. Secondly, many of the extant families are heterogeneous or intermediate for the primitive and advanced characters tabulated. Most probably the primitive character states predominated in the Cretaceous and Tertiary ancestors of these present day families. With respect to the unexpectedly high proportions of Cretaceous families with specialised rays and storied structure, one could also invoke reversibility as a phenomenon responsible for these results. For the development of heterogeneous from homogeneous rays and of nonstoried from storied structure, paedomorphosis would provide one possible mecha-nism.

In spite of all restrictions the above analysis confirms the major Baileyan trends for vessel perforation type, fibre wall pitting and parenchyma distribu-tion. There can also be no doubt about the phylogenetic derivation of the primitive vessel element from long tracheids, although the fossil record of Lower Cretaceous vesselless angiosperm wood is still a matter of discussion (Wolfe et al., 1975). There are simply no alternatives if one considers the ves-selless xylem of all potential candidates for angiosperm ancestry in the fossil record (Barghoorn, 1964) together with the occurrence of vesselless extant Angiosperms, neatly fitting at the primitive end of the Baileyan morphological series for tracheids and vessel members.

Phylogenetic wood anatomy would be greatly served by a comprehensive and critical survey of all scattered records of fossil woods. Preferably this survey should be carried out along the lines of Page's study (1979), recognising different wood anatomical types, and observing the greatest possible restraint in attaching names of extant genera or families. Such a review would not only contribute to the understanding of angiosperm phylogeny, but also give a historical dimension and test of the presumed ecological trends in xylem evolution, discussed below.

Ecological wood anatomy

'... Les élements du bois, en relation avec d'importantes fonctions physiologiques, ne peuvent échapper à l'adaption.' This opinion of Vesque (1889) explains why he found wood anatomical characters of such restricted use in taxonomy because of the possibility of wood anatomical diversity being induced by environmental factors. Vesque had arrived at his conclusion by theoretical considerations on water conduction in stems (1876) in which he already used Poiseulle's law to analyse the plant structural variables which influence the capacity to withstand drought. Narrow vessels such as occur in the Ericaceae are one possible means of restricting the transpiration stream and thus maintaining a larger 'réserve transpiratoire'. Vesque went as far as to state that the ecological conditions of a plant could be predicted from wood anatomical parameters such as vessel diameter and frequency. In experimental work (Vesque and Viet, 1881) it had been demonstrated that in the xylem of peas, fewer and smaller vessels were formed when cultivated in a humid atmosphere, and this, together with Böhm's (1879) spectacular result that willow shoots immersed in water subsequently produced a vesselless secondary xylem, had disqualified wood anatomical characters as taxonomic markers of prime importance in Vesque's opinion, 'malgré l'opinion exprimée par Monsieur Solereder.'

That studies of primarily physiological and ecological importance were considered to have a direct impact on systematic anatomy demonstrates that nineteenth century botany was a beneficially integrated discipline. Ecological wood anatomy, including primarily physiological studies of factors influencing xylogenesis, continued to attract (mostly sporadic) attention throughout the first half of the present century.

In this paper I will only deal with that part of ecological wood anatomy, which tries to study correlations of presumably genetically fixed wood anatomical features with ecological preference or requirements related to growth habit. Earlier studies on these topics have been elaborately reviewed by Carlquist (1975, 1980) and to a lesser extent by myself (1973, 1976). Although ecological factors as important co-determinants of wood anatomical diversity and specialisation have always been of some concern to comparative wood anatomists from the nineteenth century onwards (and even to Leeuwenhoek in the seventeenth century), and play some role in Bailey's discussions (1944, 1953) of the major phylogenetic trends, ecological wood anatomy, approached from a broad comparative basis, can be considered to have started its flowering period as recently as 1975 when Carlquist published his 'Ecological Strategies of Xylem Evolution'. Under this telling title, wood anatomical diversity is re-interpreted as the result of functionally adaptive evolution, with far-reaching explanations for many anatomical features to be encountered most frequently in different ecological and taxonomic categories of vascular plants. Correlations between ecology and higher incidence of certain wood anatomical traits in natural groups such as for instance the Compositae (Carlquist, 1966), between regional floras with major ecological differences (Kanehira, 1921; Baas, 1973) and studies on plant groups of specific habit (*e.g.* Gibson, 1973) are at the basis of Carlquist's approach, to which he has added a wealth of original observations, not only in his publications mainly related to ecological wood anatomy, but also in each of his numerous studies on systematic anatomy (many cited in Carlquist, 1975, 1980). Current concepts that water transport in plants is often associated with considerable negative pressures and is thus subjected to the risk of embolism (Zimmermann and Brown, 1971), speculations on the need for mechanical strength of vessels to withstand the suction forces to which they are subjected, and the necessity for wood fibres and tracheids to meet with requirements of mechanical strength for support of the crown and resistance to wind, have inspired the functional interpretations of the correlations observed. Since the precise role of any of these factors in the survival of a single plant is still largely a matter of speculative guess work in the absence of comprehensive experimental data, and since many correlations remain to be more soundly based on more taxonomic groups and woody floras with different ecologies, some of Carlquist's hypotheses have been challenged in subsequent publications (Baas, 1976, 1977; Van Vliet, 1976, 1979). Other studies have (partly) supported some of the functionally adaptive hypotheses (*e.g.*, Dickison, 1979, 1980; Dickison *et al.*, 1978; Baas, 1979). In a defence of his approach Carlquist (1980) reacts to some of his critics: 'I do feel that looking for potential functions is a more productive method of inquiry than looking for nul hypotheses or for random variation.' One can endorse these aims, but in addition it is an absolute necessity to put these hypotheses to the test of their more general, and thus predictive, value. In this respect many recent wood anatomical studies pay tribute to Carlquist's stimulating role, by examining whether certain ecological trends can be observed in various taxa. Such critical studies, together with an even more urgent need of experimental and ecological data, on a large variety of woody plants are a prerequisite for the further development of ecological and functional wood anatomy into a fully mature, scientific discipline. Uncritical acceptance of untested functional interpretations of ecological trends may lead to such curious comparisons of montane elements like *Deutzia* of the Saxifragaceae to desert shrubs with respect to vessel diameter and vessel frequency (Styer and Stern, 1979). This would qualify most cool temperate shrubs for successful desert life (Carlquist, 1980, has also criticised this comparison).

Meanwhile the analysis of ecological trends in classical wood anatomy remains much hampered by a very high 'noise level', *i.e.*, within-a-tree and random within-stand variation of most quantitative and some qualitative characters (*cf.* Akachuku and Burley, 1979; Van den Oever *et al.*, 1981) and statistical methods are required to unearth the ecological component of the variation. Carlquist's opinion (1980) that ecological trends in wood anatomy should not be expected to obey requirements for statistical significance seems a misunderstanding in this respect. Especially in wood anatomy where so many diverse factors, including non-standard sampling, influence the results, statistical tests are essential – even to discover trends which are not directly obvious in the raw data set (examples are presented by Van den Oever *et al.*, 1981).

Although exceptions remain, the following ecological trends in wood anatomy may be considered to be soundly established and of predictive value: vessel member and fibre length, and vessel diameter decrease, whilst vessel frequency increases with increasing drought or decreasing temperature. For drought, these trends have been explained by Carlquist in terms of increased safety for water conduction in xeric taxa. This is satisfactory as far as vessel diameter and frequency are concerned, but the role of decreased vessel member length (according to Carlquist another device to increase the strength of the vessel in order to withstand high negative pressures) remains obscure in my opinion. For the parallel correlations with decreasing temperatures (which rest on parallel latitudinal and altitudinal trends) such a functionally adaptive hypothesis is not satisfactory for the mostly mesic taxa to which these trends apply: even if one considers that there is some element of physiological drought in low temperatures, there is no reason to assume that there are higher demands for safety in temperate or tropical high montane trees and shrubs than in tropical lowland forests; in fact risks of embolism would be expected to be higher in the latter. A possible functional explanation would be that plants from cooler regions can afford a system which is less efficient (*i.e.*, has narrower vessels and produces more resistance to flow; increased vessel frequency hardly ever compensates sufficiently for the predominant influence of diameter) than plants in the tropical lowland with their reputed (but too rarely measured) high average transpiration rates, despite the prevailing high atmospheric humidity. The role of vessel element and fibre length also remains largely unexplained here. Decreased mechanical requirements of cool temperate representatives or of desert shrubs of small stature, easily induces the belief that these can do with shorter fibres, but in for instance *Ilex*, and notably *Symplocos,* such reasoning does not apply because the longest elements can be found in species of small trees and shrubs (Van den Oever *et al.*, 1981). The relationships between tree stature and fibre length in Dicotyledons are apparently not so straightforward as they are between tree size and tracheid length in conifers (Bannan, 1965; Carlquist, 1975), and are in need of analysis based on a large, diverse sample.

With respect to scalariform vessel perforations, floristic analyses have unambiguously shown that these are far more frequent in temperate and tropical montane floras than in the everwet tropical lowland or in arid regions (Baas, 1976); yet within widely distributed genera there appears to be no general trend with respect to temperature regimes, but only evidence of reduction of bar number in more xeric conditions (Carlquist, 1975; Dickison *et al.*, 1978; Dickison, 1979). A functional interpretation of the retention of scalariform perforations in taxa of mesic, cool environments, and their loss in habitats with higher transpiration rates, *viz.,* that there is only selection pressure for decreased resistance to flow in the latter case seems to make sense.

Ecological trends in ray size and type, parenchyma distribution and abundance, and fibre type have received less attention; largely because it is difficult to imagine differential functional advantage or disadvantage in various ecological systems.

Ray height decreases with higher latitudes and altitudes, parallel to changes in vessel member and fibre length within natural groups (Carlquist, 1966; Baas, 1973; Van der Graaff and Baas, 1974; Van den Oever *et al.*, 1981). Similar correlations hold true for increased drought in the Compositae (Carlquist, 1966). Analysis of ray type within widely distributed genera or homogeneous families has been scanty. Carlquist (1966) found that in the Compositae there is a greater tendency for procumbency in temperate than in tropical species, but in a com-

Table 2 — Incidence of major ray types in different regional floras (percentages of genera).

Region and source of data	Number of genera	Exclusively homogeneous	Nearly homogeneous or homogeneous and heterogeneous	Exclusively heterogeneous
Central Europe Schweingruber, 1978 descriptions and plates	43	16%	40%	44%
Grosser, 1977 table	52	25%	29%	46%
Japan Sudo, 1959 identification codes	111	31%	14%	45%
New Zealand Meylan & Butterfield, 1978 descriptions and plates	61	3%	18%	79%
Formosa Kanehira, 1921 descriptions	213	13%	8%	79%
South-east Asia Hayashi et al., 1973 (commercial timbers, chiefly lowland rain forest) plates only	123	22%	10%	68%
New Guinea Furuno, 1977 & 1979 (mainly lowland rain forest) plates and tables	98	14%	18%	68%
Java Moll & Janssonius, 1906-1936 (lowland and montane trees) descriptions	368	17%	5%	78%
World Metcalfe & Chalk, 1950 by family, family descriptions by species (1800)		12% 21%	21% ?	67% ?

parison of mesic and xeric species the trend is the opposite: the mesic species show a greater degree of procumbency in the ray cells. The Compositae are, however, an unfortunate example for ray analysis because of the likely occurrence of paedomorphosis in a number of its woody representatives, which complicates the issue. In order to have at least a slight idea of the existence of ecological trends in ray type, I have analysed the frequency of their occurrence in a number of detailed wood anatomical, regional surveys. These results are summarised in Table 2, and should be interpreted as a very rough guide indeed, because authors differ greatly in assigning rays to the types defined by Kribs (1935, 1968), as becomes evident from comparing Grosser's (1977) and Schweingruber's (1978) accounts of largely the same woods. The difference is probably due to the greater emphasis of Schweingruber on very low proportions of rays with some square marginal cells; nevertheless the added percentages of genera with exclusively homogeneous (*i.e.*, homocellular and composed of exclusively procumbent cells), ànd nearly homogeneous rays is fairly similar in both cases. The woody flora of Japan resembles the European one in this respect. However, in the largely temperate flora of New Zealand the percentage of genera with homogeneous rays is much lower. For the floras of Formosa, Java, and New Guinea the percentages are somewhat intermediate and quite close to those for the world flora as a whole. The opposite trend in the north temperate floras as compared to the New Zealand flora becomes especially significant if we consider that for vessel perforation type and Tertiary helical thickenings New Zealand follows the common trends for temperate and subtropical floras all over the world (Van der Graaff and Baas, 1974; Baas, 1976; Meylan and Butterfield, 1978). Further and more refined analyses are needed, but from these preliminary results I would expect that there are no preferential trends for either homogeneous or heterogeneous rays to be predominant in special climatic zones. Differences in percentages for regional floras probably merely reflect different composition with respect to taxonomic affinity. The lack of a trend for ray type is not unexpected, because functional advantage (if any) of the alternative types is presumably independent of climatic conditions.

Axial parenchyma, like ray cells, serves for storage, mobilisation and translocation of photosynthates in the wood. As emphasised by Carlquist (1975) this function can also be carried out by living fibres, and a discussion of functional and ecological trends or aspects should take both tissues into account simultaneously. At this stage, knowledge on the occurrence of living fibres is too limited (especially for tropical species) and unfortunately insufficient to fulfill this requirement. A marked feature of the tropical lowland forests is the high incidence of woods with abundant parenchyma (chiefly paratracheal) and the relative paucity of this feature in temperate floras. Table 3 substantiates this generalisation semi-quantitatively. The estimates of parenchyma abundance are purely arbitrary and were made on the basis of photomicrographs and diagrammatic drawings in the literature, together with statements in the descriptions (although the latter often had to be modified to have com-

Table 3 — Incidence of genera with various degrees of parenchyma abundance in different regional floras (percentages). See Table 2 for number of genera.

Region and source of data (see Table 2 for number of genera)	Abundant	Intermediate	Scarce
Central Europe Schweingruber, 1978 descriptions and plates	19	42	39
Grosser, 1977 descriptions and plates	19	27	54
New Zealand Meylan & Butterfield, 1978 descriptions and plates	32	30	38
Formosa Kanehira, 1921 descriptions	39	37	24
South-east Asia Hayashi et al., 1973 plates	44	45	11
Tropical West Africa Normand & Paquis, 1976 plates (215 genera)	57	21	22
Java Moll & Janssonius, 1906-1938 drawings and descriptions	44	27	29

parable measures). Though each individual wood anatomist will come up with different estimates and percentages, he will find the same trend. Further and more exact analyses of parenchyma abundance, type of distribution, and replacement by living (often septate) fibres are necessary, before a functional interpretation can even start. At this stage the trend can only surprise us, because there seems to be little sense in creating an optimal system for photosynthate storage and mobilisation in evergreen tropical trees capable of continuous photosynthesis throughout the year.

Table 4 summarises frequency of genera with unspecialised or specialised fibres (*i.e.*, fibre-tracheids and libriform fibres) in various floras. In temperate floras fibre-tracheids appear to be of more common occurence than in the tropics. And within the tropical flora of Java, suitable for analysis,

Table 4 — Incidence of genera with fibre-tracheids in different regional floras. See Table 2 for number of genera.

Region and source of data	Percentage of genera with fibre-tracheids
Central Europe Grosser, 1977	42
Japan Sudo, 1959	31
New Zealand Meylan & Butterfield, 1978	34
Java Moll & Janssonius, 1906-1938	18
World Metcalfe & Chalk, 1950	33*

* Metcalfe & Chalk give this percentage for *species* with 'distinctly bordered pits' in fibres. Possibly this also includes a number of species with fibres intermediate between libriform fibres and fibre-tracheids (*i.e.,* with small but distinctly bordered pits mainly confined to the radial walls).

most species with fibre-tracheids are typically montane elements. This trend has as parallel in mesic versus more xeric floras (Novruzova, 1968): fibre-tracheids are more common in mesic habitats. Within *Ilex* the latitudinal trend runs parallel (Baas, 1973). Whether there is any functional value for cool mesic ecologies to favour fibre-tracheids, or rather, why libriform fibres should be more suitable in hot dry òr everwet conditions is unclear, unless one invokes the correlated phenomenon of vessel perforation specialisation, which seems a sensible adaptation to minimise resistance to flow in either hot ànd/or dry conditions. It is, however, far from evident why fibres should specialise simultaneously with vessel perforations, and there are many examples where this is not the case (*e.g.* numerous Rubiaceae and Myrtaceae with fibre-tracheids but simple perforations).

It remains, in my opinion, a hazardous excercise to try and force all trends observed into functional models. In a recent paper (Van den Oever *et al.,* 1981) we drew attention to the very strong mutual correlations between most quantitative variables in the wood of *Symplocos*, which are of course what one should expect in an integrated, complex tissue like wood. If such correlations are inherent to the genetic basis for xylem differentiation, changes favourable for one function (and possibly of critical significance in the evolutionary success of the offspring) might induce other changes which have no function or even function less perfectly than the ancestral character state. This phenomenon could be manifest as functionless variation in the eye of the beholder, but would in fact have originated as a by-product of adaptation. I do not abandon the idea of a fair proportion of *'patio ludens'* (*cf.* Van Steenis, 1976) *i.e.,* random variation in wood structure without a special survival value, but wish to add correlative phenomena as a complementary explanation of functionless traits. In the zoological literature, Gould and Lewontin (1979) have given a most apt example of how the observer can be misled when biassed in favour of functional interpretations: in the San Marco cathedral in Venice the four spandrels (triangular spaces at the intersection of the main arches) below the central dome seem to be the starting point of design. Yet they are merely by-products of a building scheme for a dome resting on four arches, and their triangular shape is dictated by architectural restraint. If we transpose this example to functionally explained, ecological wood anatomy, a skeptical mind might argue that in the various, often conflicting functions which wood has to carry out, we cannot even tell the spandrels from the arches. In a more positive, but as yet more speculative vein, we may guess that in dicotyledonous trees vessel diameter and type of perforation, possibly together with fibre length and fibre wall thickness are the arches, and that some other wood structural variables could be spandrels. The former might hold the greatest promise for functional interpretation; the latter for taxonomic markers.

Towards a synthesis

In the previous sections on systematic, phylogenetic, and ecological wood anatomy, cross references from one to the other could not be avoided for the simple reason that they are inseparable aspects of the study of wood evolution: its present 'end' products, course and possible mechanisms. Of course, individual studies in systematic and ecological wood anatomy can have other, legitimate goals. Also, ecological wood anatomy can only shed light on some adaptive aspects of wood evolution, and does not take into account other, perhaps major pathways and factors involved such as saltatory changes (not necessarily adaptive), or historical and correlative restraints for certain characters (*cf.* Van Steenis, 1976, 1981, and Gould, 1980, for evolutionary models diverging from the Neo-Darwinistic one of gradual accumulation of adaptive changes).

Each aspect, systematic, phylogenetic, or ecological (including functional), is of profound significance for the other, as already realised by nineteenth century botanists. It is for instance obvious that paleobotanists will consider the present-day ecological preferences of certain wood anatomical types in their interpretations of the fossil record (as does Page, 1979, 1980). Similarly the work of a systematic anatomist is more rewarding if he attempts to contribute to a phylogenetic classification with an awareness of ecological factors that may have played a role in the evolution of the wood anatomical diversity in the taxa studied. Fortunately this synthetic approach is becoming more and more common in modern studies, but there is still scope for more integration and faster progress. A research program for the near future should therefore give high priority to the following aspects:

1. Continued critical studies on wood anatomical variation within clearly defined taxonomic groups (families, genera, species) of both wide ànd narrow ecological amplitudes in order to refine and expand our knowledge on general or group specific ecological trends, ànd on wood anatomical variation which is nòt related to ecological preference. Attention should also be paid to differences in growth habit within the taxa analysed.

2. Comprehensive surveys of the wood anatomy in restricted floras with diverse vegetation types. Such studies should take detailed climatic and edaphic factors into account (cf. Carlquist, 1977) and would allow a refinement of the very rough floristic analyses carried out so far. Emphasis should be on representative sampling of all woody species. Such field-based studies are especially required for tropical floras in regions with great local differences in elevation, rainfall and soils. Meanwhile there is still a wealth of descriptive information to be tapped from the existing wood anatomical, floristic, and ecological literature for the search of possibly major ecological trends. Simultaneously, attention should be focussed on the total range of wood structural variation within florulas of a more or less constant ecology.

3. A critical review of the fossil record of gymnosperm and angiosperm wood in order to add historical dimensions, wherever possible to phylogenetic and ecological trends in wood anatomy. This great task should preferably be planned as a cooperative venture of paleobotanists and specialists on the identification of extant woods with the backing of representative reference wood and slide collections.

4. Studies to improve our understanding of the genetic control and potential of phenotypic variation of both quantitative and qualitative wood anatomical variation, and of correlative phenomena operative in the control of xylem differentiation.

5. A more broad-based knowledge on the limiting factors (if any) in the functioning of wood in various ecological conditions, both as a conduc-

tive and mechanical tissue. Only tree physiologists and technologists can provide such knowledge, and all we can ask for is to include a greater (wood anatomical) diversity of species, especially from the tropics, in their experimental studies.

6. Integration of anatomical and physiological data on wood with information from other plant parts, especially roots and leaves. This is especially relevant for the functional interpretation of wood as a water conducting system, because it may well be that the limiting factors for this function reside in root or leaf structure.

Most of these needs, of equal priority, have been recognised by others (notably by Carlquist). Yet, I have listed them here, in order to stress the feasibility of progress on all aspects in the forthcoming years. This is because all priorities listed also have a great potential for economically important applications of the study of wood structure and function.

The international cooperation advocated for a new system of computerised wood identification (Miller, 1980; IAWA Committee, 1981) would profit from the same descriptive data as are needed to comply with the first three priorities listed above. Detailed regional wood anatomical surveys are moreover of national significance, and their execution would be a highly commendable activity in several small research centres for wood science, because progress and success can be achieved without recourse to expensive equipment or extensive library facilities. Likewise, improved understanding of genetic control of wood structure and of limiting factors in its functioning are of significance in the applied forestry and forest products sciences.

In conclusion, the quest for the Holy Grail (*i.e.*, the study of wood evolution) can continue and be stimulated, irrespective of whether research programs are judged and funded on purely scientific criteria or on potential economic value. All we need is an increased awareness of the complex problems which all students of wood can contribute to in their own way.

Acknowledgements

I am much indebted to Dr. J. Muller (Rijksherbarium) for giving access to his extensive data on the fossil pollen record prior to publication, and for useful advice and discussions. Prof. Dr. C.G.G.J. van Steenis critically read the manuscript and suggested some useful improvements.

54

References

Akachuku, A.E. & J. Burley. 1979. Variation of wood anatomy of Gmelina arborea Roxb. in Nigerian pantations. IAWA Bull. 1979/4: 94-99.

Arber, A. 1941a. Tercentenary of Nehemiah Grew (1614-1712). Nature 147: 630-632.

Arber, A. 1941b. Nehemiah Grew and Marcello Malpighi. Proc. Linn. Soc. London 153: 218-238.

Armstrong, J.E. & D.T. Funk. 1980. Genetic variation in the wood of Fraxinus americana. Wood and Fiber 12: 112-120.

Baas, P. 1969. Comparative anatomy of Platanus kerrii Gagnep. Bot. J. Linn. Soc. 62: 413-421.

Baas, P. 1973. The wood anatomy of Ilex (Aquifoliaceae) and its ecological and phylogenetic significance. Blumea 21: 193-258.

Baas, P. 1976. Some functional and adaptive aspects of vessel member morphology. In: P. Baas, A.J. Bolton & D.M. Catling (eds.), Wood Structure in Biological and Technological Research: 157-181. Leiden Bot. Series No. 3. Leiden Univ. Press, The Hague.

Baas, P. 1977. The peculiar wood structure of Leptospermum crassipes Lehm. (Myrtaceae). IAWA Bull. 1977/2: 25-30.

Baas, P. 1979. The peculiar wood structure of Vaccinium lucidum (Bl.) Miq. (Ericaceae). IAWA Bull. 1979/1: 11-16.

Baas, P. 1982. Leeuwenhoek's contributions to wood anatomy and his ideas on sap transport in plants. In: L.C. Palm & H.A.M. Snelders (eds.), Antoni van Leeuwenhoek 1632-1982. Studies Commemorating the 350th Anniversary of his Birth. Rodopi, Amsterdam.

Baas, P. & E. Werker. 1981. A new record of vestured pits in Cistaceae. IAWA Bull. n.s. 2: 41-42.

Baas, P. & R.C.V.J. Zweypfenning. 1979. Wood anatomy of the Lythraceae. Acta Bot. Neerl. 28: 117-155.

Baas Becking, L.M.G. 1924. Antoni van Leeuwenhoek, immortal dilettant. Scientific Monthly, New York 18: 547-554.

Bailey, I.W. 1933. The cambium and its derivative tissues. VIII. Structure, distribution, and diagnostic significance of vestured pits in Dicotyledons. J. Arn. Arbor. 14: 259-273.

Bailey, I.W. 1944. The development of vessels in angiosperms and its significance in morphological research. Amer. J. Bot. 31: 421-428.

Bailey, I.W. 1953. Evolution of the tracheary tissue of land plants. Amer. J. Bot. 40: 4-8.

Bailey, I.W. & W.W. Tupper. 1918. Size variation in tracheary cells. I. A comparison between the secondary xylems of vascular cryptograms, gymnosperms and angiosperms. Proc. Amer. Arts & Sci. 54: 149-204.

Bannan, M.W. 1965. The length, tangential diameter and length/width ratio of conifer tracheids. Canad. J. Bot. 43: 967-984.

Barghoorn, E.S. 1964. Evolution of cambium in geological time. In: M.H. Zimmermann (ed.), The Formation of Wood in Forest Trees: 3-17. Acad. Press, New York/London.

Bary, A. de. 1877. Vergleichende Anatomie der Vegetationsorgane der Phanerogamen und Farne. Englemann, Leipzig.

Boerlage, J.G. 1875. Bijdrage tot de kennis der houtanatomie. Thesis, Leiden.

Böhm, J. 1879. Über die Function der vegetabilischen Gefässe. Bot. Ztng. 37: 214-255.

Brazier, J. 1975. The changing pattern of research in wood anatomy. J. Microscopy 104: 53-64.

Bruyne, A.S. de. 1952. Wood structure and age. Proc. Koninkl. Nederl. Akad. Wet. Amsterdam, Series C, 55: 282-286.

Candolle, A.P. de. 1818. Systema naturale regni vegetabilis. Treuttel & Würtz, Paris.

Carlquist, S. 1961. Comparative plant anatomy. Holt, Rinehart & Winston, New York.

Carlquist, S. 1962. A theory of paedomorphosis in dicotyledonous woods. Phytomorphology 12: 30-45.

Carlquist, S. 1966. Wood anatomy of Compositae: a summary, with comments on factors controlling wood evolution. Aliso 6: 25-44.

Carlquist, S. 1970. Wood anatomy of Hawaiian, Macaronesian, and other species of Euphorbia. Bot. J. Linn. Soc. Suppl. 60, Vol. 63 (1): 181-193.

Carlquist, S. 1975. Ecological strategies of xylem evolution. Univ. California Press, Berkeley/Los Angeles/London.

Carlquist, S. 1977. Ecological factors in wood evolution: a floristic approach. Amer. J. Bot. 64: 887-896.

Carlquist, S. 1980. Further concepts in ecological wood anatomy, with comments on recent work in wood anatomy and evolution. Aliso 9: 499-553.

Chalk, L. 1937. The phylogenetic value of certain anatomical features of dicotyledonous woods. Ann. Bot. n.s. 1: 409-428.

Chalk, L. & C.R. Metcalfe. 1979. The history of wood anatomy. In: C.R. Metcalfe & L. Chalk (eds.), Anatomy of the Dicotyledons 1: 4-9 (2nd Ed.). Oxford Univ. Press, Oxford.

Chalon, J. 1868. Anatomie comparée des tiges ligneuse dicotylédonées. Deuxième Mémoire, Gand. Seen in: Bull. Soc. bot. France 16, revue bibl.: 38-39.

Chattaway, M.M. 1955. Crystals in woody tissues I. Trop. Woods 102: 55-74.

Chattaway, M.M. 1956. Crystals in woody tissues II. Trop. Woods 104: 100-124.

Cronquist, A. 1968. The evolution and classification of flowering plants. Nelson, London.

Dahlgren, R.M.T. 1980. A revised system of classification of the angiosperms. Bot. J. Linn. Soc. 80: 91-124.

Dickison, W.C. 1975. The bases of angiosperm phylogeny: vegetative anatomy. Ann. Missouri Bot. Gard. 62: 590-620.

Dickison, W.C. 1979. A note on the wood anatomy of Dillenia (Dilleniaceae). IAWA Bull. 1979/2-3: 57-60.

Dickison, W.C. 1980. Comparative wood anatomy and evolution of the Cunoniaceae. Allertonia 2: 281-321.

Dickison, W.C., P.M. Rury & G.L. Stebbins. 1978. Xylem anatomy of Hibbertia (Dilleniaceae) in relation to ecology and evolution. J. Arn. Arbor. 59: 32-49.

Dobell, C. 1932. Antony van Leeuwenhoek and his 'little animals'. Bale & Danielson, London.

Duhamel du Monceau, H.L. 1758. La physique des arbres. Paris.

Engler, A. & K. Prantl. 1889-1897. Die natürlichen Pflanzenfamilien 1-4. Englemann, Leipzig.

Furuno, T. 1977. Anatomy of Papua New Guinea woods. Research Report of Foreign Wood (Shimane Univ.) No. 6.

Furuno, T, 1979. *Ibidem* (continued). *Ibidem,* No. 8.

Gibson, A.C. 1973. Wood anatomy of the Cactoideae (Cactaceae). Biotropica 5: 29-65.

Gould, S.J. 1980. Is a new general theory of evolution emerging? Paleobiology 6: 119-130.

Gould, S.J. & R.C. Lewontin. 1979. The spandrels of San Marco and the Panglossian paradigm: a critique of the adaptionist programme. Proc. R. Soc. London B 205: 581-598.

Graaff, N.A. van der, & P. Baas. 1974. Wood anatomical variation in relation to latitude and altitude. Blumea 22: 101-121.

Gregory, M. 1980. Wood identification: an annotated bibliography. IAWA Bull. n.s. 1: 3-41 (also separately: IAWA, Leiden).

Grew, N. 1682. The anatomy of plants with an idea of a philosophical history of plants. Rawlins, London.

Grosser, D. 1977. Die Hölzer Mitteleuropas. Springer, Berlin/Heidelberg/New York.

Hartig, T. 1859. Beiträge zur vergleichenden Anatomie der Holzpflanzen. Bot. Ztng. 11: 93-101; 105-112.

Hayashi, S., T. Kishima, L.C. Lau, T.M. Wong & P.K. Balan Menon. 1973. Micrographic atlas of Southeast Asian timbers. Wood Research Institute, Kyoto.

Heniger, J. 1973. Antoni van Leeuwenhoek. In: C. Coulston Gillispie (ed.), Dictionary of Scientific Biography 8: 126-130. Scribner's, New York.

Hill, J. 1770. The construction of timber (from its early growth explained by the microscope and proved from experiments in a great variety of kinds). R. Baldwin, London.

Hogeweg, P. & J. Koek-Noorman. 1975. Wood anatomical classification using iterative character weighing. Acta Bot. Neerl. 24: 269-283.

Hooke, R. 1665. Micrographia: or some physiological descriptions of minute bodies made by magnifying glasses with observations and inquiries thereupon. Martyn & Allestry, London (repr. 1961, Cramer, Weinheim).

IAWA Committee. 1981. Standard list of characters suitable for computerized hardwood identification (with explanation of coding procedure and characters by R.B. Miller). IAWA Bull. n.s. 2: 99-145 (also separately: IAWA, Leiden).

Jane, F.W. 1963. Botanical aspects of wood science. In: W.B. Turrill (ed.), Vistas in Botany 2: 1-35. Pergamon, London/New York/Paris/Los Angeles.

Kanehira, R. 1921. Anatomical characters and identification of Formosan woods. Govt. of Formosa, Taihoku.

Kieser, D.G. 1815. Elemente der Phytonomie 1. Phytotomie. Cröcker, Jena.

Koek-Noorman, J. 1980. Wood anatomy and classification of Henriquezia Spruce, Platycarpum Humb. et Bonpl. and Glaesonia Standl. Acta Bot. Neerl. 29: 117-126.

Koek-Noorman, J. & P. Hogeweg. 1974. The wood anatomy of Vanguerieae, Cinchoneae, Condamineae and Rondeletieae (Rubiaceae). Acta Bot. Neerl. 23: 627-653.

Koek-Noorman, J., P. Hogeweg, W.H.M. van Maanen & B.J.H. ter Welle. 1979. Wood anatomy of the Blakeae (Melastomataceae). Acta Bot. Neerl. 28: 21-43.

Kribs, D.A. 1935. Salient lines of structural specialization in the wood rays of dicotyledons. Bot. Gaz. 96: 547-557.

Kribs, D.A. 1968. Commercial foreign woods on the American market. Dover Publications, New York.

Leeuwenberg, A.J.M. (ed.). 1980. Loganiaceae. In: A. Engler & K. Prantl (contd by P. Hiepko & H. Melchior), Die natürlichen Pflanzenfamilien, 2nd Ed., 28 b I (especially: Anatomy of the secondary xylem by A.M.W. Mennega, p. 112-161).

Leeuwenhoek, A. van. 1673-1696 (1939-1982). The collected letters I-XI (to be continued). Swets & Zeitlinger, Amsterdam.

Mabberley, D.J. 1974. Pachycauly, vessel-elements, islands and the evolution of arborescence in 'herbaceous' families. New Phytol. 73: 977-984.

Mägdefrau, K. 1973. Geschichte der Botanik. Fischer, Stuttgart.

Malpighi, M. 1686. Opera omnia. London (repr. 1687, Leiden).

Marum, M. van. 1773. Diss. inauguralis philosophica de motu fluidorum in plantis. Thesis, Groningen, The Netherlands.

Metcalfe, C.R. 1959. A vista in plant anatomy. In: W.B. Turrill (ed.), Vistas in Botany 1: 76-99. Pergamon, London/New York/Paris/Los Angeles.

Metcalfe, C.R. 1979. History of systematic anatomy. In: C.R. Metcalfe & L. Chalk (eds.), Anatomy of the Dicotyledons 1: 1-4 (2nd Ed.). Oxford Univ. Press, Oxford.

Metcalfe, C.R. & L. Chalk. 1950. Anatomy of the Dicotyledons 1 & 2. Clarendon Press, Oxford.

Meylan, B.A. & B.G. Butterfield. 1978. The structure of New Zealand woods. DSIR Bull. 222. Wellington.

Miller, R.B. 1977. Vestured pits in Boraginaceae. IAWA Bull. 1977/3: 43-48.

Miller, R.B. 1980. Wood identification via computer. IAWA Bull. n.s. 1: 154-160.

Möbius, M. 1901. Marcellus Malpighi: Die Anatomie der Pflanzen. Englemann, Leipzig.

Moll, J.W. & H.H. Janssonius. 1906-1938. Mikrographie des Holzes der auf Java vorkommen-den Baumarten 1-6. Brill, Leiden.

Möller, J. 1876. Beiträge zur vergleichenden Anatomie des Holzes. Vienna, Gerold (abstracted in French in Bull. Soc. Bot. France 24: 3).

Muller, J. 1981. Fossil pollen records of extant Angiosperms. Bot. Review 47: 1-142.

Normand, D. & J. Paquis. 1976. Manuel d'identification des bois commerciaux. 2. Afrique guinéo-congolaise. CTFT, Nogent s/Marne.

Novruzova, Z.A. 1968. The water-conducting system of trees and shrubs in relation to ecology. Baku, Izd. AN. Azerb. SSR (in Russian).

Oever, L. van den, P. Baas & M. Zandee. 1981. Comparative wood anatomy of Symplocos and latitude and altitude of provenance. IAWA Bull. n.s. 2: 3-24.

Page, V.M. 1979. Dicotyledonous wood from the Upper Cretaceous of central California. J. Arn. Arbor. 60: 323-349.

Page, V.M. 1980. *Ibidem.* II. J. Arn. Arbor. 61: 723-748.

Robbertse, P.J., G. Venter & H. Janse van Rensburg. 1980. The wood anatomy of South African Acacias. IAWA Bull. n.s. 1: 93-103.

Sachs, J. 1875. Geschichte der Botanik. Oldenbourg, München.

Sanio, C. 1863. Vergleichende Untersuchungen über die Zusammensetzung des Holzkörpers. Bot. Ztng. 21: 359-363; 369-375; 377-385; 389-399; 401-412.

Schierbeek, A. 1950 & 1951. Antoni van Leeuwenhoek – zijn leven en werken. De Tijdstroom, Lochem, The Netherlands.

Schierbeek, A. 1959. Measuring the invisible world. The life and works of Antoni van Leeuwen-hoek FRS. Abelard-Schuman, London/New York.

Schmucker, T. & G. Linnemann. 1951. Geschichte der Anatomie des Holzes. In: H. Freund (ed.), Handbuch der Mikroskopie in der Technik V/1: 3-77. Umschau, Frankfurt.

Schweingruber, F.H. 1978. Mikroskopische Holzanatomie. Formenspektren mitteleuropäischer Stamm- und Zwerghölzer zur Bestimmung von rezentem und subfossilem Material. Kommis-sionsverlag, Zürich.

Solereder, H. 1885. Ueber den systematischen Wert der Holzstructur bei den Dicotyledonen. Thesis, München.

Solereder, H. 1899 & 1908. Systematische Anatomie der Dicotyledonen & Ergänzungsband. Enke, Stuttgart.

Steenis, C.G.G.J. van. 1976. Autonomous evolution in plants. Differences in plant and animal evolution. Gardens' Bull. (Singapore) 29: 103-126.

Steenis, C.G.G.J. van. 1981. Rheophytes of the world. An account of flood-resistant flowering plants and ferns and the theory of autonomous evolution. Sijthoff & Noordhoff, Alphen a/d Rijn, The Netherlands.

Stern, W.L. 1978. A retrospective view of comparative anatomy, phylogeny, and plant taxo-nomy. IAWA Bull. 1978/2-3: 33-39.

Stern, W.L. 1982. Highlights in the early history of the International Association of Wood Anat-omists. This volume.

Styer, C.H. & W.L. Stern. 1979. Comparative anatomy and systematics of woody Saxifragaceae. Deutzia. Bot. J. Linn. Soc. 79: 291-319.

58

Sudo, S. 1959. Identification of Japanese hardwoods. Bull. Govt. Exper. Station No. 118: 1-138.

Takhtajan, A.L. 1980. Outline of the classification of flowering plants (Magnoliophyta). Bot. Rev. 46: 225-359.

Thorne, R.F. 1976. A phylogenetic classification of the Angiospermae. Evolutionary Biology 9: 35-106.

Timell, T.E. 1980. Karl Gustav Sanio and the first scientific description of compression wood. IAWA Bull. n.s. 1: 147-153.

Vesque, J. 1876. Recherches anatomiques et physiologiques sur la structure du bois. Ann. Sci. nat. 6e sér. bot. 3: 358-371.

Vesque, J. 1881. De l'anatomie des tissus appliquée à la classification des plantes. Nouv. Arch. Muséum (Paris) 2e sér. 4: 1-56.

Vesque, J. 1889. De l'emploi des caractères anatomique dans la classification des végétaux. Bull. Soc. Bot. France 36: 41-87.

Vesque, J. & C. Viet. 1881. Influence du milieu sur les végétaux. Ann. Sci. nat. 6e sér. bot. 12: 165-176.

Vliet, G.J.C.M. van. 1976. Wood anatomy of the Rhizophoraceae. In: P. Baas, A.J. Bolton & D.M. Catling (eds.), Wood Structure in Biological and Technological Research: 20-75. Leiden Bot. Series No. 3. Leiden Univ. Press, The Hague.

Vliet, G.J.C.M. van. 1979. Wood anatomy of the Combretaceae. Blumea 25: 141-223.

Vliet, G.J.C.M. van. 1981. Wood anatomy of the Old World Melastomataceae. Blumea 27: 395-462.

Werker, E. & P. Baas. 1981. Trabeculae of Sanio in secondary tissues of Inula viscosa (L.) Desf. and Salvia fruticosa Mill. IAWA Bull. n.s. 2: 69-76.

Wolfe, J.A., J.A. Doyle & V.M. Page. 1975. The bases of angiosperm phylogeny: paleobotany. Ann. Missouri Bot. Gard. 62: 801-824.

Zimmermann, M.H. 1979. The discovery of tylose formation by a Viennese Lady in 1845. IAWA Bull. 1979/2-3: 51-56.

Zimmermann, M.H. & C.L. Brown. 1971. Trees - structure and function. Springer, Berlin/New York.

Functional xylem anatomy of angiosperm trees

MARTIN H. ZIMMERMANN

Harvard Forest, Petersham, Massachusetts 01366, U.S.A.

Summary: Water appears to exist in the xylem of trees under negative pressures for long periods of time, often for years. At the same time, it can move through wood in axial direction with relative ease. This remarkable combination of efficiency and safety is only possible because of the intricate three-dimensional structure of wood. Small-diameter and short vessels are safer water conductors, wide and long ones are more efficient. This paper describes how and where vessels end, how water moves from one vessel to the next, and how length distribution and the three-dimensional network of vessels make wood an efficient and safe water conductor.

Introduction

One of the major functions of wood is the conduction of water and soil nutrients from roots to leaves. The cohesion theory which explains this process, has been doubted for many years on the basis of two previously unanswered questions. The first of these was whether negative pressures are universally present in the xylem of terrestrial plants under transpirational conditions. The introduction of the pressure bomb by Scholander *et al.* (1965) has made measurements of negative pressures and pressure gradients possible. The results of such measurements have been quite consistent with the predictions of the cohesion theory.

The second reason why the cohesion theory was doubted was the question how water could possibly exist in the xylem for long periods of time (in some cases years) under negative pressure, *i.e.* in a metastable condition. This question has received very little attention. Stability is obviously the result of water being trapped in very small compartments (tracheids or vessels). The key to successful conduction under these circumstances is an arrangement of compartments (tracheids and vessels) in the tree which lets water move easily from compartment to the next, thus providing for safe and efficient long-distance transport.

This chapter is specifically concerned with the three-dimensional arrangement of vessels within the tree, a topic that has received little attention in the past, yet is of very considerable functional significance.

The conducting units: vessels

Safety vs. efficiency

Vessel width and length are parameters that determine the efficiency and safety of water conduction. The wider and longer vessels are, the higher is their conductivity (or lower the resistance to flow). In ideal capillaries, conductivity is proportional to the fourth power of the radius (or diameter) (Zimmermann, 1978b). This means that, at a given pressure gradient, the relative volumes of water flowing through capillaries of diameters 1, 2, and 4, are 1, 16, and 256 respectively. Vessels differ from ideal capillaries in that their walls are not absolutely smooth and that they are of finite length. Water must periodically pass from one vessel to the next. This represents a certain additional flow resistance. Several authors have tried to estimate this by comparing measured volume flow rate through a piece of wood with volume flow rate calculated from measured tracheary element diameters. For coniferous wood measured conductivities were around 30% of the theoretical one, for dicotyledons, of the order of 50% (see the citations in Zimmermann and Brown, 1971). Results of such measurements are only approximate, probably because it is very difficult to obtain precise vessel diameter measurements. The fourth-power relationship causes a diameter error of 5% to become a volume flow error of 22%. For similar considerations, we may consider resistance to flow from vessel to vessel unimportant. For purposes of calculating resistance to flow, we can regard vessels with 50% efficiency like endless capillaries that are 15% narrower.

Vessel diameter varies so widely in different wood species that its effect upon conductivity is far greater than the effect of vessel length. Trees with narrow vessels, such as our common diffuse-porous species, have diameters of the order of 50 to 100 μm, while trees with wide vessels, such as our ring-porous species, have diameters of 200 to 500 μm.

Let us now consider how efficiency is related to safety. A wide-vessel tree like one of our deciduous (ring-porous) oaks *(Quercus)* has large earlywood vessels with a diameter of *c.* 300 μm. We know that these are quite vulnerable and are lost from the conducting system during the course of the winter. Every spring, oaks must produce a new set of large earlywood vessels before the new leaves unfold (Huber, 1935; Zimmermann and Brown, 1971). Because these new vessels are the first wood produced, they are arranged in a ring, hence the name ring-porous. Their large diameter makes them such efficient water conductors, that a single ring is sufficient to conduct all the water needed by the crown. This is the high efficiency, high risk principle (Huber, 1935). It has its drawbacks, because certain diseases affecting the outer wood (even if only peripherally as in the chestnut blight) can be devastating (Huber, 1956; Zimmermann and McDonough, 1978).

A narrow-vessel tree like one of our maples *(Acer)* has vessel diameters of about 75 μm. The vessels of oak are four times wider and about 30 times longer (see the section below) than those of maple. In order to carry the same amount of water at a given pressure gradient, maple needs about 7000 times as many vessels as does oak. From this we see immediately how much more vulnerable a wide-vessel tree is. Visualise that a single vessel is lost by some accident such as a nibble by a bark beetle. The damage in oak is 7000 times more serious than in maple. There is also evidence that the tensile strength of water is greater in smaller compartments than in larger ones. This question has not been systematically investigated, but the literature shows that the greatest tensile strength of water has been measured in the smallest compartments, the cells of fern annuli (Greenidge, 1957).

Vessel length

Inside vessel diameter is easy to see and record. Vessel length however, is a much more difficult parameter to measure. It has been known for years that vessel diameter and length are correlated: wide vessels are longer than narrow ones. It has also been known for many years that air can be blown through a piece of stem only if vessels have been cut open at both ends. This means that the longest vessels of a species are slightly longer than the longest piece of stem through which air can be forced (*e.g.* Handley,1936; Greenidge, 1952). But it was not until the publication of the paper by Skene and Balodis (1968) that we acquired the concept of vessel-length distribution.

Vessel-length distribution can be measured by perfusing the stem with very small paint particles. The injected particles move only as far as the vessel ends, because they are too large to cross the vessel-to-vessel pit membranes. By lateral loss of water (filtration), they accumulate until vessels are packed with them and can be easily recognised with a stereo microscope. The piece of wood is cut into equal segments and the paint-containing vessels are counted at different distances from the point of injection. Assuming random distribution of vessels in the wood, one can calculate distribution of vessel lengths. Instead of paint, one can use air-flow-rate measurements at given pressure gradients.

62

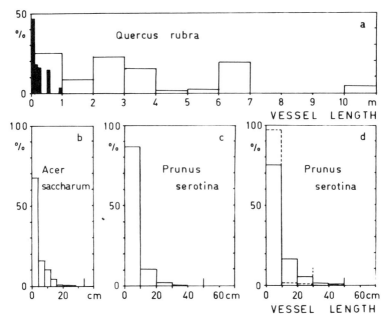

Fig. 1. Vessel-length distribution in the stems of three tree species. Note that the percentage scale is the same in all diagrams (a-d), but the vessel-length scale in (a) is different from those in (b-d) (Recalculated and redrawn from Zimmermann and Jeje, 1981). – a. Vessel-length distribution in *Quercus rubra*. The one-metre wide bars show the length classes of the wide earlywood vessels, the narrow black bars are narrow latewood vessels. – b. In *Acer saccharum* vessels are not only narrower than in *Quercus*, but also much shorter. The longest vessel length is indicated by the small vertical bar at 34 cm. There are so few vessels in the longer length classes (*e.g.* 32-34 cm) that the bars often do not exceed the thickness of the baseline of the graph. – c. Length distribution in *Prunus serotina* is comparable to that in *Acer*. – d. Vessel-length distribution in *Prunus serotina* shown separately for earlywood (solid lines) and latewood (dashed lines).

These methods work quite well with woods containing large numbers of relatively short vessels, because random distribution is no problem. However, if some vessels are of a length comparable to stem length, they cannot possibly be randomly distributed in the stem because there is not enough room. Under these circumstances, the calculations become less reliable (Zimmermann and Jeje, 1981).

Figure 1 shows two typical distribution patterns. Earlywood vessels of oak (*Quercus rubra*, Fig. 1a) with diameters of the order of 300 μm are up to 11 metres long. Among these wide vessels there are also shorter ones, 0.5 to several metres in length. As examples of narrow-vessel wood, those of sugar maple (*Acer saccharum*) and black cherry (*Prunus serotina*) are given in Fig. 1b, c and d. Note that the percentage scales of all diagrams are the same, but that the vessel length scales are different. The longest vessels of the narrow-vessel woods are only about 0.5 metres long. Interesting is the fact that the latewood vessels of oak (shown as narrow, black bars) are of a width and length comparable to those of a narrow-vessel tree.

The most important single result of these vessel-length distribution measurements is the fact that only very few vessels (often only a fraction of a percent) are of the longest length class. By far the largest percentage of the vessels is very short (Zimmermann and Jeje, 1981). This is important for two reasons, a 'historical' and a functional one. With few exceptions, all earlier papers dealing with vessel length, report about longest lengths, although the authors do not usually say so. Indeed they might often not have been aware of the fact that they dealt only with a tiny fraction of all vessels. There may be a developmental reason why there are more short than long vessels. If vessel ends are randomly produced, long vessels should be rarer than short ones. Functionally, the short vessels are of extreme importance for the safety of the tree. This is evident in a much more dramatic manner, if we look at the original vessel counts. In the case of the cherry tree shown in Fig. 1c, 29,307 vessels were cut and infused with paint. At a distance of 10 cm from the infusion, 2038 vessels (7%) carried paint, at 20 cm it was 281 (1%), at 30 cm 42, and at 40 cm 6 vessels. It is indeed very important for the safety of water conduction in cherry that 10 cm from an injury 93% of the vessels are uninjured.

Vessel ends and vessel-to-vessel pits

Individual terminal vessel elements have been described in the literature (*e.g.* Bierhorst and Zamora, 1965; Handley, 1936). They can be found in macerated wood. It is much more difficult to see them in sections. In herbaceous plants vessel ends have occasionally been demonstrated by India ink injections. This is very unreliable, because India ink particles are large, ragged and clog a vessel easily as water filters out laterally (Zimmermann and Jeje, 1981) (anyone who has tried to fill an ordinary fountain pen with India ink will agree!). Vessel ends have remained elusive for many years. It was not until P.B. Tomlinson and I introduced our cinematographic method of analysis in 1965 that vessel ends became visible (for a description of these methods see Zimmermann, 1976). We can now 'travel' through the wood and observe the path of vessels, their beginnings and endings. Such a film is commercially available (Zimmermann, 1971). Cinematographic analysis has shown that vessels end mostly in pairs or groups. This makes sense functionally, because water must move from one vessel to the next via vessel-to-vessel pits. Narrow vessels have also been reported to end in contact with rays (Bosshard *et al.*, 1978). But these are not of hydraulically significant size for long-distance transport.

Where vessels run in pairs or groups, their common walls are beautifully structured to permit the passage of water from one vessel to the next, because the secondary cell walls arch over much of the primary wall pair, exposing a relatively large pit-membrane area, at the same time providing mechanical support. The importance of these vessel-to-vessel pit areas cannot be overemphasised. Very short segments of vessel-to-vessel pit areas are illustrated dramatically on scanning electron micrographs. But their full extent can be shown only with three-dimensional vessel-network reconstructions from motion-picture film (*e.g.* Fig. IV-6 in Zimmermann and Brown, 1971). One has to keep in mind that such illustrations, which show the reconstructed vessel network, are usually about 10 times foreshortened. Vessel overlap areas are therefore some 10 times longer than they appear in the illustration.

Hydraulic architecture of trees

Vessel distribution within the tree

Plant anatomists have known for a long time that the vessels of the xylem near the leaves are more 'primitive', not only in reference to the structure of the perforation plate, but also narrower, than vessels farther away from the leaves. Vessel diameters of a tree stem increase in diameter continuously as one moves down the stem, away from the crown (*e.g.* Fig. 6 in Zimmermann, 1978a). Vessel diameters are larger in some roots than in the stem (Riedl, 1937). We have recently measured vessel-length distribution in different parts of *Acer rubrum* and found that the same relationship holds for vessel length (Zimmermann and Potter, 1982). It was, of course, to be expected, because length is correlated with diameter. Figure 2 shows how vessel diameter is related to longest vessel length and to the percentage of vessels in the shortest length class (0-4 cm).

Correlation of vessel length and diameter is also evident within single growth rings. Figure 1a shows this most dramatically for the case of the ring-porous *Quercus*. The wide earlywood vessels range in length from near one to about eleven metres, while all the narrow latewood vessels are shorter than one metre. In a semi-ring-porous species like *Prunus serotina,* the same phenomenon is found (Fig. 1d). The longest vessels of the earlywood are in the 40-50 cm length class, those of the latewood in the 20-30 cm class. At the same time, latewood (the dashed lines in Fig. 1d) contains a larger percentage of vessels in the shortest class (0-10 cm). We have found even in genera which are considered diffuse-porous, like *Acer,* that the latewood vessels are not only narrower, but measurably shorter.

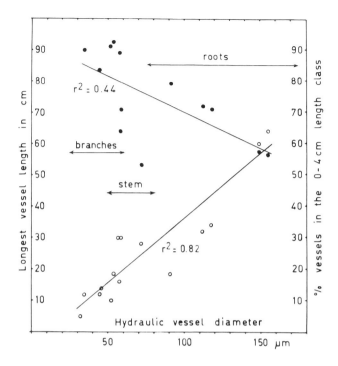

Fig. 2. Vessel-length distribution in branches, stem, and roots of *Acer rubrum*. This graph shows the correlation of hydraulic vessel diameter and vessel length. The wider the vessels, the longer the longest vessels (○) and the fewer vessels in the shortest length class of 0-4 cm (●) (From Zimmermann and Potter, 1982).

Vessel network and pattern of sap ascent.

We have known for some time that when we inject a dye into a radial drill hole in the tree stem, the dye will ascend (and of course descend also because of the artificial pressure-gradient reversal below the injection hole), and spread in circumferential direction within the growth ring along its path away from the injection point. Cinematographic analysis has shown us that this is because vessels do not run strictly parallel within a growth ring, but cross over each other slightly. Depending upon tree species this is more or less pronounced. Even 'straight-grained' types, such as ash *(Fraxinus),* show such a pattern (Zimmermann, 1971). This means that as we follow the axial path of water up the tree, it fans out tangentially.

In addition to the spreading of the vessel path, each ring shows a certain amount of 'spiral grain' which may differ considerably from one ring to the next (Fig. IV-5 in Zimmermann and Brown, 1971). In other words, the path of water is not straight up the tree, it follows a rather complex pattern. Functionally this means that a single root does not supply a single branch, but a rather large section of the crown. Or, a single branch is not supplied by a single root, but by a large part of the root system. This must have considerable benefits if part of the root system or crown is damaged.

Leaf-specific conductivity

Let us now look at the movement of water up the tree as a whole. I like to refer to xylem anatomy of the tree as a whole as the hydraulic architecture. From the tree's point of view, the transverse section of any part of the stem, branch, or twig, is used to transport the water to all leaves above that point. To my knowledge, Huber (1928) was the first person to recognise the importance of this way of looking at water conduction. He related the transverse-sectional area of xylem to the fresh weight of leaves that are supplied by that part of the xylem. I like to refer to this as the Huber value. It is generally about 0.5 square millimetres per gram fresh weight of supplied leaves for aerial axes of our north temperate trees.

Huber (1928) made an important discovery while making such measurements. He found that the leader of a small fir tree was greatly favoured with xylem supply tissue, when compared to lateral branches. This is an interesting expression of apical dominance. Huber (1928) worked with small conifers where transverse-sectional xylem area is a useful measure. As soon as we look at older trees that have significant amounts of non-conducting xylem, and if we want to compare angiosperm types in which vessel sizes vary widely, xylem transverse-sectional area becomes meaningless, because conductivity depends very much more upon vessel diameter than upon xylem transverse sectional area. We can greatly improve Huber's method by measuring flow rates of water through stem or branch pieces, at a defined pressure gradient, and express this per leaf quantity. I like to call this leaf-specific conductivity. Physiologically, it would perhaps be best to express conductivity per leaf surface area, a unit which is meaningful also in respect to photosynthesis. But it is rather awkward to measure in a relatively large tree. We therefore use fresh weight as Huber (1928) did. Fresh weight is − for our purposes at least − a very stable value. Thick leaves are heavier per surface area than thin leaves, even if we look at leaves from different parts of the same tree. But one can nevertheless find a reasonable conversion factor for 'average leaves', if such a conversion is desirable for a comparison of different tree species.

So far we have published leaf-specific conductivities (LSC) only for some North American diffuse-porous tree species (Zimmermann, 1978a). A sample

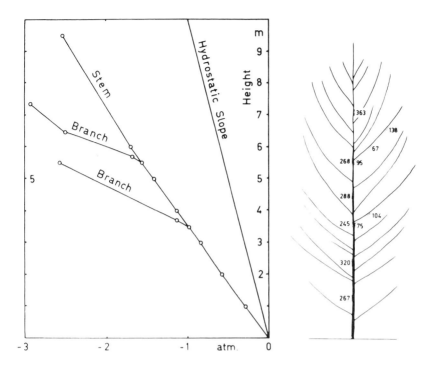

Fig. 3. Leaf-specific conductivity (LSC) in *Populus grandidentata*. The tree is illustrated to scale on the right. LSC are given at different locations. Conductivities in the stem are about twice those of the branches. The graph on the left has been calculated from measured LSC, and is based upon two assumptions, first that the pressure at the base of the tree is zero. If this were not the case, the graph would merely have to be shifted to the left (or right). Second, it is assumed that all leaves transpire fully and equally. The hydraulic architecture may be a means to regulate transpiration within the crown (see text) (Diagram on right from Zimmermann, 1978a).

is given in Fig. 3 on the right, illustrating *Populus*. The numbers have been arrived at by a somewhat arbitrary choice of dimensions in order to make LSC cover a range of about 1-1000. It may be near 1 at the petiole insertion, and nearly 1000 in the stem of *Betula*. The number is the flow rate in microlitres per hour, per gram fresh weight of supplied leaves, at the pressure gradient of gravity (0.1 atm./m).

Figure 3 shows a tree of about 9.5 metres height, LSC values are given for 15 cm long pieces taken from points along the stem and branches. First of all, we see that the conductivity of the stem is about twice as great as that of the branches. This means that from the flow resistance point of view, it is easier to get the water up the stem than out into a branch. The higher conductivity of the stem compensates (partly) for the disadvantage of the greater height of the top leaves to which water has to be lifted. If we assume a uniform transpiration rate for all leaves, we can calculate the pressure gradients along stem and branches from LSC numbers. This is shown in the graph on the left half of Fig. 3. Under conditions of no flow, such as in the early morning before sunrise, we should find a hydrostatic gradient ('hydrostatic slope' in Fig. 3). As water movement begins in the morning, pressures become more and more negative and approach the pressure graph on the left, which has been calculated for full transpiration. This graph is based upon two assumptions. The first assumption is that the pressure is zero at the base of the stem. If this is not the case, the curve simply has to be shifted to the left (or perhaps slightly to the right), but it would not change its shape. The second assumption is that all leaves transpire at the same rate. This may not be the case. Indeed, we must assume that this is how hydraulic architecture regulates transpiration: as the pressure in lateral branches (and its leaves) drops (becomes more negative), transpiration may be curtailed. Hydraulic architecture is a structural means of favouring leaves at the top of the tree. The leaves at the top of the tree are the most important ones for the tree's survival, but they are in the most unfavourable position: water has to be brought over the longest distances to them and has to be lifted to the greatest heights. This disadvantage is overcome by the overall xylem structure. Larger vessels provide a low resistance path to the top of the tree, smaller vessels provide a high resistance path to lower leaves. Lower leaves can therefore not take water at the expense of the top leaves.

Hydraulic segmentation extends into the peripheral parts of the tree. LSC is lower in twigs than in branches, and lower in petioles than in twigs (Zimmermann, 1978a). LSC distribution appears to be a very important safety device in another respect. If the tree experiences a very severe drought during which some parts are permanently damaged, it is extremely important that the most valuable parts are saved. For example, it would be disastrous for the tree, if the stem xylem would suffer extensive cavitation. This does not appear to happen. The lowest pressures are experienced in the leaves. If pressures drop to a lethal level, they will reach it in the leaves first; leaves will wilt and die. If stresses become still greater, individual twigs or branches may be lost by cavitation. But it is of paramount importance to the survival of the tree that the stem is saved, because it represents many years of investment and could not easily be replaced. This is particularly important for palms which cannot repair stem tissue by secondary growth (Zimmermann, 1978b; Zimmermann et al., 1982). We might go a step further and speculate, as P.B. Tomlinson (pers. comm.) suggested, that the success of this construction caused trees with lateral branches to survive during evolution, and trees with dichotomous branching to become extinct.

A feature that is often encountered is a hydraulic bottleneck (*i.e.* a decrease of vessel diameters resulting in a particularly low LSC) at junctions from stem to branch, branch to twig, and twig to petiole (Larson and Isebrands, 1978; Zimmermann, 1978a) This causes a sharp pressure drop across the junction. When we discovered this phenomenon we considered it to be an important part of the pressure gradient from roots to leaves. However, we do not always find a significant bottleneck. When we made plots like the one shown in Fig. 3 left, we realised that the bottleneck does not always significantly change the pressure distally, because it occurs only over a very short distance. Its significance may be quite a different one. Vessels at the branch junction are conspicuously narrower than either above or below (Fig. 7 in Zimmermann, 1978a; Fig. 2 in Larson and Isebrands, 1978). At the same time, the Huber value is usually somewhat larger, that is, the xylem transverse sectional area shows a slight bulge. The function of the bottleneck is therefore probably not only to induce a particularly sharp pressure drop, but also to represent a safety barrier: here the water columns will not break under excessive stress, they will break beyond, the vessels are again wider (*i.e.* less safe).

In conclusion, let us compare LSC and the Huber value. LSC vary widely throughout the tree and obviously give us information about hydraulic architecture. The Huber value, on the other hand, appears to be quite constant throughout many tree species and seems to be an indication of the mechanical requirements of the tree.

References

Bierhorst, D.W. & P.M. Zamora. 1965. Primary xylem elements and element associations of angiosperms. Amer. J. Bot. 52: 657-710.

Bosshard, H.H., L. Kučera & U. Stocker. 1978. Gewebe-Verknüpfungen in Quercus robur L. Schweiz. Zeit. Forstwesen 129: 219-242.

Greenidge, K.N.H. 1952. An approach to the study of vessel length in hardwood species. Amer. J. Bot. 39: 570-574.

Greenidge, K.N.H. 1957. Ascent of sap. Ann. Rev. Plant Physiol. 8: 237-256.

Handley, W.R.C. 1936. Some observations on the problem of vessel length determination in woody dicotyledons. New Phytol. 35: 456-471.

Huber, B. 1928. Weitere quantitative Untersuchungen über das Wasserleitungssystem der Pflanzen. Jahrb. Wiss. Bot. 67: 877-959.

Huber, B. 1935. Die physiologische Bedeutung der Ring- und Zerstreutporigkeit. (Physiological significance of ring- and diffuse-porousness). Ber. Dtsch. Bot. Ges. 53: 711-719. (Xerox copy of English translation available from National Translation Center, 35 West 33rd St., Chicago, IL 60616, U.S.A.)

Huber, B. 1956. Die Gefässleitung. In: O. Stocker (ed.), Encyclopedia of Plant Physiology. Vol. 3: 541-582. Springer, Berlin/Göttingen/Heidelberg.

Larson, P.R. & J.G. Isebrands. 1978. Functional significance of the nodal constricted zone in Populus deltoides Bartr. Canad. J. Bot. 56: 801-804.

Riedl, H. 1937. Bau und Leistungen des Wurzelholzes. (Structure and function of root wood). Jb. wiss. Bot. 85: 1-75. (Xerox copy of English translation available from National Translation Center, 35 West 33rd St., Chicago, IL 60616, U.S.A.)

Scholander, P.F., H.T. Hammel, E.D. Bradstreet & E.A. Hemmingson. 1965. Sap pressures in vascular plants. Science 148: 339-346.

Skene, D.S. & V. Balodis. 1968. A study of vessel length in Eucalyptus obliqua L'Hérit. J. Exp. Bot. 19: 825-830.

Zimmermann, M.H. 1971. Dicotyledonous wood structure made apparent by sequential sections. (Film E 1735. Film data and summary available as a reprint. Inst. f. d. wiss. Film, Nonnenstieg 72, 34 Göttingen, West Germany.)

Zimmermann, M.H. 1976. The study of vascular patterns in higher plants. In: I.F. Wardlaw & J.B. Passioura (eds.), Transport and Transfer Processes in Plants: 221-235. Acad. Press, New York.

Zimmermann, M.H. 1978a. Hydraulic architecture of some diffuse-porous trees. Canad. J. Bot. 56: 2286-2295.

Zimmermann, M.H. 1978b. Structural requirements for optimal water conduction in tree stems. In: P.B. Tomlinson & M.H. Zimmermann (eds.), Tropical Trees as Living Systems: 517-532. Cambridge Univ. Press. London/New York/Melbourne.

Zimmermann, M.H. & C.L. Brown. 1971. Trees: structure and function. Springer, New York/ Heidelberg/Berlin.

Zimmermann, M.H. & A. Jeje. 1981. Vessel-length distribution in stems of some American woody plants. Canad. J. Bot. 59: 1882-1892.

Zimmermann, M.H., K.F. McCue & J.S. Sperry. 1982. Anatomy of the palm Rhapis excelsa. VIII. Vessel network and vessel-length distribution in the stem. J. Arn. Arbor. 63: 83-95.

Zimmermann, M.H. & J. McDonough. 1978. Dysfunction in the flow of food. In: J.G. Horsfall & E.B. Cowling (eds.), Plant Disease. An Advanced Treatise. Vol. 3: 117-140. Acad. Press, New York.

Zimmermann, M.H. & D. Potter. 1982. Vessel-length distributions in branches, stem, and roots of Acer rubrum L. IAWA Bull. (in preparation).

Cell wall hydrolysis in the tracheary elements of the secondary xylem

B.G. BUTTERFIELD and B.A. MEYLAN

Botany Department, University of Canterbury, Christchurch, New Zealand, and Physics and Engineering Laboratory DSIR, Lower Hutt, New Zealand

Summary: Tracheid and vessel element pit membranes and vessel perforation plate partitions are reduced by hydrolysis to a cellulose residue during cell differentiation. This process removes the non-cellulosic matrix components from unlignified walls leaving a cellulose microfibrillar web or net. Where these webs traverse small openings as in tracheid and vessel element pit membranes, they usually survive intact. Where they traverse larger openings as in vessel element perforations, they generally disappear. Whether this is due to the greater forces of the transpiration stream across such openings or the activity of cellulase is discussed.

Secondary xylem fulfills two main roles in plant stems and roots: namely conduction of water and dissolved ions, and structural support for the plant body. In coniferous woods, the greater bulk of the axial cell system is made up of tracheids which fulfill both of these functions. Long distance transport in the xylem in these woods is accomplished by water passing from tracheid to tracheid through the intertracheary pit-pairs (Fig. 1). With certain exceptions *(e.g. Phyllocladus),* these pits are confined to the radial walls of the tracheids, except the last few tracheids of the latewood and the first tracheids of the earlywood where they may also occur on the tangential walls. In earlywood tracheids the secondary wall is raised to form a border that overarches the pit membrane and pit chamber. The development and microfibrillar structure of these borders has been described by Harada and Côté (1967). The pit membrane is circular in outline and during differentiation behind the cambium the periphery is enzymatically digested (Fengel, 1972) to open up a network of radiating cellulose microfibrils (the margo) subtending a central undigested region (the torus) (Fig. 2). In some woods the matrix components have also been digested from the torus but the microfibrils remain closely packed (Fig. 3); in other woods the torus is undigested thereby retaining its matrix components; while in still other woods, notably *Pinus* (Fig. 4), *Tsuga* and *Cedrus*, the torus may be thickened by the apposition of additional substances (Liese, 1965). In most cases the matrix substances of the margo are totally removed, and a comparison of the cellulose content between the torus and margo suggests that some of the cellulose in the margo may be removed also (Fengel, 1972).

Tracheid to tracheid water flow in coniferous earlywood, therefore, follows a pathway through the pit aperture of one cell into the pit chamber, through the margo of the unaspirated pit membrane and out through the pit chamber and aperture of the pit-pair into the adjacent cell (Fig. 1). Water flow may be more restricted in the latewood where the pit membrane margo remains undigested. This would be in keeping with the transition in tracheid function in some coniferous woods from primarily conduction in the earlywood to primarily support in the latewood (Butterfield and Meylan, 1980).

Under certain conditions of stress, such as air embolism by drying, the conifer intertracheid pit membrane aspirates to one side of the pit chamber effectively sealing the pit-pair (Petty, 1971; Gregory and Petty, 1973; Bolton and Petty, 1977). Pit aspiration is made possible by the extension of the microfibrillar net of the margo and occurs as a natural phenomenon in heartwood and at sites of injury. Latewood pit membranes, however, are less prone to aspiration than earlywood ones, thus accounting for the higher permeability values sometimes recorded for dry latewood (Siau, 1971). Thomas (1967) has shown in southern pines that the latewood pit membranes have a denser margo than earlywood pit membranes due to a substantially higher content of margo microfibrils. Thomas and Kringstad (1971) suggested that this fact, together with a lower pit membrane diameter, increased the strength of the pit membranes in latewood and presumably accounts for the lower frequency of pit aspiration in the latewood.

Long distance water flow in most hardwoods occurs through vessels, which are essentially long axial conduits (Fig. 5). They differentiate behind the cambium from individual vessel elements that develop open pores grouped into perforation plates in their common walls. Perforation plates normally develop in the end walls but not exclusively so. All types of perforation plate (reticulate, scalariform, simple, intermediate, and combination or dimorphic forms) differentiate by the enzymatic degradation of the perforation partitions consisting of primary walls and middle lamella (Butterfield and Meylan, 1972; Meylan and Butterfield, 1972a, 1981a; Esau and Charvat, 1978; Murmanis, 1978; Benayoun et al., 1981) (Fig. 6). There is more or less universal agreement that this process involves the enzymatic removal of the non-cellulosic matrix components (Kuster, 1956; Niedermeyer, 1974; Ohtani and Ishida, 1976; Murmanis 1978; Esau and Charvat, 1978; Meylan and Butterfield , 1981a; Benayoun et al., 1981). However, the process that is responsible for the removal of the cellulosic residue traversing each aperture in a perfora-

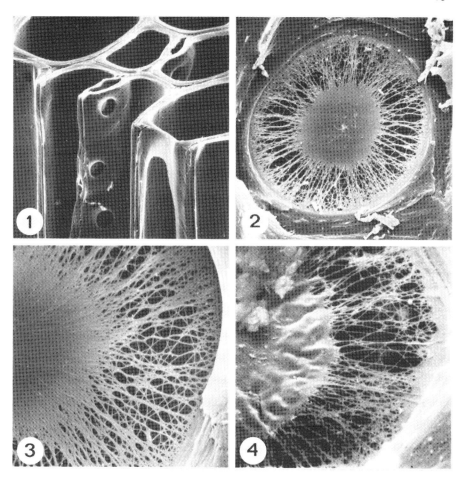

Fig. 1. Transverse and tangential longitudinal faces of *Pinus radiata* D. Don. The imperforate tracheids that constitute most of the axial cell system are interconnected by bordered pit-pairs mostly on the radial walls. × 700. – Fig. 2. An intertracheid pit membrane in *Phyllocladus glaucus* Carr. exposed by wall fracture. The margo has been hydrolysed leaving a network of radiating cellulose macrofibrils subtending a central torus. × 3100. – Fig. 3. Detail of an intertracheid pit membrane in *Dacrydium cupressinum* Lamb. The matrix components are digested from both the torus and the margo and there is no clear boundary between these areas. × 8300. – Fig. 4. Detail of an intertracheid pit membrane in *Pinus radiata* D. Don. The margo is thickened by an additional deposition or incrustation and is accordingly protected from enzymatic attack. The boundary between the torus and margo is more sharply defined than in *Dacrydium*. × 6400.

tion plate is still the subject of much debate. The observation of fine microfibrillar webs traversing each opening, particularly in scalariform perforation plates (Ishida 1970; Bonner, 1970; Meyer and Muhammad, 1971; Meylan and Butterfield, 1972b, 1981a; Butterfield and Meylan, 1972; Parham, 1973; Thomas and Bonner, 1974; Ohtani and Ishida, 1976, 1978a) (Fig. 7), and occasionally in simple perforation plates (Butterfield and Meylan, 1980) has led to a belief that the cellulosic residue is swept away by the water flow following the onset of the transpiration stream (Yata *et al.*, 1970; O'Brien, 1974, 1981). Others believe that the cellulosic as well as the matrix components of the perforation partitions are enzymatically digested (Sassen, 1965; Niedermeyer, 1974; Ohtani and Ishida, 1976; Murmanis, 1978). Cellulase activity has been detected in xylem sap (Sheldrake, 1970) but its role in perforation plate development remains unverified.

Microfibrillar webs in mature wood are generally confined to the lateral extremities of openings close to the end of long scalariform plates and sometimes traverse the entire opening of the last few perforations. Webbed openings often grade into normal scalariform pit membranes at each end of the perforation plate. In the scalariform perforation plates of *Liriodendron*, Thomas and Bonner (1974) observed webs only in the smaller openings (5-8 μm) and not in the larger openings (13-20 μm). All these observations suggest that the strength of the residual webs is dependent on the diameter of the openings that they traverse. This would support the concept that microfibrillar webs traversing large openings are not strong enough to survive the transpiration stream.

Microfibrillar webs traversing simple perforation plates have been observed only very rarely (Meylan and Butterfield, 1981a). This led us to suggest that such a stage in the breakdown of the partitions in simple perforation plates must exist even if only briefly but that its chance of survival is very small because of the large size of the opening it traverses.

More recently, however, we have observed several different stages of perforation plate development in the same vessel. In these cases some partitions between vessel elements were still intact, some were partly digested and others were completely digested. Since full water flow clearly could not have commenced in such vessels these observations raise doubts that the cellulose component survives the enzymatic digestion to be subsequently removed by the transpiration stream. A more plausible argument, at least in the case of simply perforated vessels, could be that both the non-cellulosic and the cellulosic components are enzymatically attacked and that in some cells, autolysis occurs *before* the cellulose component has been completely digested. If microfibrillar webs are left, then they may be removed later by the transpiration stream. This would account for the occasional observation of residual microfibrils hanging out from between the overarching borders in some of the simple

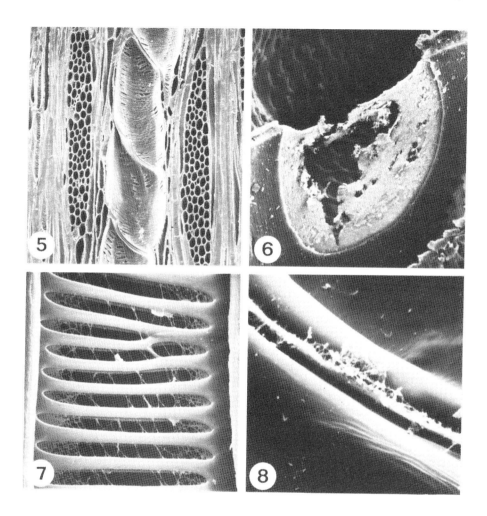

Fig. 5. Tangential longitudinal face of *Beilschmiedia tawa* (A. Cunn.) Benth. & Hook. *f. ex* Kirk. Water in hardwoods flows virtually unobstructed from cell to cell through the perforation plates in the vessel elements end walls. × 140. – Fig. 6. A differentiating simple perforation plate in *Knightia excelsa* R. Br. Following hydrolysis of the matrix components, the cellulose residue is removed by unknown processes. × 1150. – Fig. 7. Cellulose microfibrils surviving in a fully differentiated scalariform perforation plate of *Quintinia acutifolia* Kirk. The position of these webs has led to speculation that, following hydrolysis of the matrix components, the cellulose residue is swept away by the transpiration stream. × 1900. – Fig. 8. Detail of the rim of a fully differentiated simple perforation plate in *Knightia excelsa* R. Br. Some cellulose microfibrils remain in the position of the digested perforation plate partition. × 5600.

perforation plates in some woods (Fig. 8 and see illustrations in Ohtani and Ishida, 1978a and Meylan and Butterfield, 1981a).

Benayoun *et al.* (1981) have suggested that the problem of cellulose removal at sites of developing perforation plates presents no problem in *Populus* as the perforation partitions contain little or no cellulose. This result contrasts strikingly with the observation of cellulose webs in differentiating perforation plates widely reported elsewhere. Benayoun *et al.* (1981) have also suggested that the last stages of the perforation plate breakdown may differ with the species or with 'xylem age': end wall breakdown beginning in the centre in carnation, along the rim in barley (Sassen, 1965) or in several places as in oak (Murmanis, 1978) and *Knightia* (Meylan and Butterfield, 1972a). From our scanning electron microscope observations we doubt if a consistent pattern exists in any one particular species. In *Knightia,* for example, we have observed perforation plates disintegrating around the rim, near the centre, and over their entire surfaces.

Intervessel pits in hardwoods lack the raised border common to the inter-tracheary pits in most coniferous earlywood tracheids. Their pit apertures vary in outline from circular to lenticular to almost slit-like in those species with close helical thickenings in their vessels (Butterfield and Meylan, 1980). Their pit membranes are circular and generally lack a distinct torus, though tori have been noted in a few species (Ohtani and Ishida, 1978b). The most significant contribution to our knowledge of the intervessel pit membrane remains that of Schmid and Machado (1968). Working with various leguminous species, these authors noted that during differentiation the intervessel pit membranes swelled, lost their encrusting substances and when viewed in the transmission electron microscope became very transparent to electrons. Vessel to parenchyma pit membranes showed a similar but asymmetrical degradation of the matrix substances, with the modifications to the pit membrane being largely confined to the vessel element side.

Essentially similar results have since been reported by Yata *et al.* (1970) and O'Brien (1970, 1974, 1981). The latter author has used the term 'hydrolysis' when referring to the enzymatic degradation of the primary part of both the end and side walls of vessels because he believed the non-cellulosic polysaccharides were hydrolytically removed (O'Brien and Thimann, 1967; O'Brien, 1970). Where the tracheary element is adjacent to a parenchyma cell the tracheary element side of the pit membrane is also hydrolysed but the digestion process stops at the middle lamella region with the extent of hydrolysis varying both between pit membranes within the one cell and also between adjacent cells (O'Brien, 1981). By O'Brien's definition, pit membranes in vessels are 'hydrolysed walls' (O'Brien, 1974).

The 'hydrolysis' of pit membranes in vessel elements is not just a side effect occurring during perforation plate development but must be necessary for long distance transport. The vessel, although many times longer than a coniferous tracheid, nonetheless has ultimately imperforate ends. This means that the flow of water from one vessel to another must occur through the intervessel pits for it to move further than the average vessel length up the trunk of the tree.

More recently the intertracheid pit membranes in the tracheids of vessel-less woods have been studied by Meylan and Butterfield (1981b). Vessel-less secondary xylem is confined to the members of the Winteraceae, Tetracentraceae, Trochodendraceae, Amborellaceae and *Sarcandra* of the Chloranthaceae (Carlquist, 1975). Although they resemble many other hardwoods in having multiseriate rays, these woods have only tracheids and axial parenchyma cells in their axial cell system. Water flow therefore is assumed to occur in the imperforate tracheids. In *Pseudowintera* the intertracheid pit-pairs are bordered and distributed more or less evenly along the radial walls (Fig. 9), though some tangential pitting occurs in the latewood. The intertracheary pit membranes are circular in surface outline and show a loose, open reticulate microfibrillar texture characteristic of hydrolysed walls (Figs. 10 and 12). No torus is present nor do the microfibrils show any tendency towards radial alignment as in the margo of conifers. The tracheid to ray and axial parenchyma pit membranes show considerable variability in the degree of hydrolysis. In cross section, all show hydrolysis of the tracheid side but the extent of hydrolysis of the parenchyma side varies. Most show the usual asymmetric half-hydrolysed pit membrane (Fig. 11) but some have the entire pit membrane hydrolysed. Parenchyma to parenchyma pit membranes in *Pseudowintera* show no evidence of hydrolysis. Instead they show a dense even texture pierced by small groups of plasmodesmata. These results are again consistent with the concept of an enzymatic release of hydrolases from the differentiating tracheid prior to the death of its protoplast. The non-cellulosic matrix components of the pit membranes are digested leaving a porous membrane presumably capable of permitting sufficient tracheid to tracheid water movement to satisfy the requirements of long distance transport in the tree.

The timing and mechanism of the attack on perforation plate partitions and pit membranes in tracheary elements remains unclear. Several authors have noted a swelling of the perforation partitions (Esau and Hewitt, 1940; Buvat, 1964; Sassen, 1965; Meylan and Butterfield, 1972; Niedermeyer, 1974; Murmanis, 1978; Esau and Charvat, 1978) and pit membranes (Schmid and Machado, 1968) prior to their hydrolysis. In the case of perforation plate partitions, Buvat (1964), Sassen (1965), Murmanis (1978) and Benayoun *et al.* (1981)

78

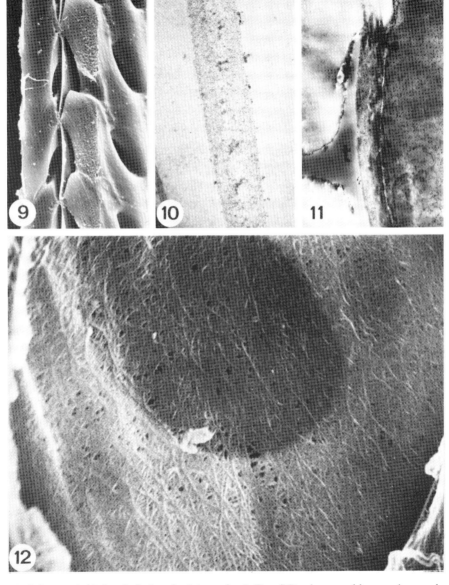

Fig. 9. Intertracheid pit-pairs in *Pseudowintera colorata* (Raoul) Dandy, a vessel-less wood exposed by a tangential cut. Note the apparently solid featureless pit membranes. × 2800. – Fig. 10. Intertracheid pit membrane in *Pseudowintera colorata*. In section the pit membranes are electron transparent, a feature characteristic of hydrolysed walls. × 29,000. – Fig. 11. Tracheid to axial parenchyma pit membrane in *Pseudowintera colorata*. Showing the asymmetric hydrolysis of the membrane to the tracheid side (t). × 18,000. – Fig. 12. Surface view of an intertracheid pit membrane in *Pseudowintera colorata* showing the open microfibrillar texture of the hydrolysed fully differentiated membrane. The darker area is the outline of the pit aperture of the tracheid behind the membrane. × 26,000.

believe this to be due to the deposition of additional wall material on the surface of the differentiating perforation partition. Esau and Charvat (1978) suggested that the swelling could be due to the removal of some of the matrix materials by wall degrading enzymes and the consequent uptake of water by the remaining carbohydrates. The labelling work of Picket-Heaps (1967) tends to support this theory suggesting that the swelling of the partition marks an early activity of wall degrading hydrolases that have attacked the wall prior to protoplast collapse. Certainly our scanning electron micrographs of perforation plate development (Meylan and Butterfield, 1972a, 1981a; Butterfield and Meylan, 1972, 1980), while illustrating the swelling phenomenon as a stage in partition breakdown, show no evidence of the deposition of any additional wall material.

Unfortunately, to date, the only published study on intertracheary pit membrane development that specifically records their swelling during differentiation is that of Schmid and Machado (1968). Further investigation may reveal that this stage is common to both perforation plate partition and pit membrane differentiation.

The timing of the enzymatic attack on partitions and pit membranes with respect to the death of the protoplast has been the subject of speculation. Early opinion held that the hydrolases were able to begin their attack on these cell wall areas only after the final collapse of the cell protoplast (O'Brien, 1970). The observations of Esau and Charvat (1978) that the swelling phenomenon occurs while the tracheary elements still possess an active protoplast tends to suggest that the cytoplasm plays an active part in cell wall hydrolysis (O'Brien, 1981). The principle difficulty in ascertaining this point is that few partly-degenerated protoplasts have ever been observed in differentiating tracheary elements. This is either because the process of cell autolysis is very rapid or, as suggested by O'Brien (1981), the macromolecular units in partly degenerated tracheary protoplasts are difficult to stabilise by fixation.

It is similarly not clear why some parts of the cell wall are hydrolysed and others are not. The basic composition of the perforation plate partitions and pit membranes prior to their hydrolysis is structurally similar, *i.e.*, they are three layered structures built up of the primary walls of the two adjoining cells and the intervening middle lamella. They differ chemically from the primary walls and middle lamella that subtend them in that they are unlignified (Bamber, 1961). This has led to the general belief that lignin protects all the other areas of the cell wall from enzymatic attack (O'Brien, 1970). This protection is particularly noticeable in multiple perforation plates where the lignified secondary wall dividing bars are not affected by the enzymes that attack the perforation partitions between them. Protection by lignin, however, does not adequately explain why the margo region of coniferous intertracheary pit membranes are attacked but the torus is not. Depositions and incrustations on the central region of the pit membrane could explain the selectivity of the attack in some cases. Most coniferous species, however, do not have thickened tori. This problem led O'Brien and Thimann (1967) to speculate that a polyuronide enrichment of the torus may protect if from hydrolysis.

Equally curious is the partial or asymmetric digestion of pit membranes between tracheary cells and axial and ray parenchyma cells. Vessel-associated parenchyma cells sometimes possess an additional wall layer overlying the normal primary and secondary walls and pit membranes. This layer was termed the 'protective layer' by Schmid (1965) since she believed it played a role in preventing the hydrolases released by the tracheary element from damaging the protoplast of the living parenchyma cell. The layer was also thought to play a role in tylose formation (Foster, 1967; Meyer, 1967; Meyer and Côté, 1968; Murmanis, 1975, 1976). Following rupture of the pit membrane the protective layer grows into the vessel lumen forming the tylose wall.

Chafe and Chauret (1974) subsequently reported the existence of an 'isotropic layer' in axial and ray parenchyma cells in a number of hardwoods including some that did not form tyloses. This layer was present in cells not in direct contact with the vessels as well as in vessel-associated parenchyma. It was found to be similar to a primary wall, rich in lignin and pectic substances but poor in cellulose. From the work of Fuji *et al.* (1979, 1980), however, it is clear that the protective layer can also occur in species that do not form tyloses. The difficulty in separating the protective and isotropic layers on histochemical grounds has led Fuji *et al.* (1981) to favour the term 'amorphous layer' for all such formations. These authors have noted that the amorphous layer is rich in hemicelluloses and contains some pectic substances and cellulose microfibrils. When first deposited the layer is poor in lignin but becomes 'lignin-rich' at a later stage of development.

When lignified and overlying the pit membranes in axial and ray parenchyma cells adjoining tracheary elements, it is reasonable to accept that the amorphous layer may perform a 'protective' function. However, in many hardwoods the amorphous layer does not overlie the pit membrane on the parenchyma cell side and a more serious objection to its protective role lies in the fact that it is not present at all in many species (Czaninski, 1979). It is also absent in some vessel-less woods (Meylan and Butterfield, 1981b) and has never been reported in the parenchyma cells of conifers (Czaninski, 1977). In the absence of any evidence to the contrary, one might conclude that hydrolysis of the pit membrane proceeds from the tracheary cell side in a tracheid or vessel element to parenchyma cell pit-pair and stops for reasons unknown at some point in the region of the middle lamella or slightly beyond it. Such a sequence, however, is difficult to reconcile with the results of Esau and Charvat (1978) and Benayoun *et al.* (1981), who have all suggested that wall hydrolysis begins at the middle lamella and progresses outwards to the primary wall layers in the breakdown of the perforation plate partition. If their observations are correct and also apply to pit membranes then it is likely that the protection from enzymatic attack is actually present in the pit membrane itself but asymmetrically distributed to the parenchyma cell side. Hydrolysis of the membrane would then begin in the middle lamella region but in most pit membranes could only proceed completely towards the tracheary element side of the pit membrane.

The extent to which the parenchyma to vessel pit membrane is hydrolysed may well be a factor in determining whether tyloses develop or not. Assuming that the matrix components act as bonding agents in the cell wall, fully hydrolysed pit membranes would be more susceptible to rupture than partly hydrolysed ones thereby allowing the tylose forming layer to grow through the pit aperture. The importance of the strength factor of the residual cellulose microfibrils in hydrolysed pits has already been demonstrated indirectly by Chattaway (1949). Based on data from more that 1100 genera she was able to demonstrate that tyloses were more likely to develop in woods whose vessel to parenchyma pit apertures exceed 10 μm than in woods with smaller ones. Although Chattaway did not relate this correlation to pit membrane ultrastructure, it is again quite reasonable to assume that the residual cellulose microfibrils of the hydrolysed pit membranes are more likely to separate when stretched over larger openings than over smaller ones.

In summary, tracheid and vessel element pit membranes and vessel perforation plate partitions are reduced by hydrolysis to a cellulose residue in the form of microfibrillar webs. This process, in effect, removes the non-cellulosic matrix components from unlignified walls. Where these microfibrillar webs traverse small openings, as in tracheid and vessel element pit membranes, they survive intact. Where they traverse larger openings as in vessel element perforation plate partitions they generally disappear. Whether this is due to the greater forces of the transpiration stream across such openings as suggested by many, or whether cellulase activity is concentrated in such areas removing the cellulose residue as well, remains a matter of speculation.

82

References

Bamber, R.K. 1961. Staining reaction of the pit membrane in wood cells. Nature 191: 409-410.

Benayoun, J., A.M. Catesson & Y. Czaninski. 1981. A cytochemical study of differentiation and breakdown of vessel end walls. Ann. Bot. 47: 687-698.

Bolton, A.J. & J.A. Petty. 1977. Variation in susceptibility to aspiration of bordered pits in conifer wood. J. Exp. Bot. 28: 935-941.

Bonner, L.D. 1970. Ultrastructure and polymer composition of intercellular passageways in yellow poplar. PhD Diss., North Carolina State Univ., N.C.

Butterfield, B.G. & B.A. Meylan. 1972. Scalariform perforation plate development in Laurelia novae-zelandiae A. Cunn.: A scanning electron microscope study. Austr. J. Bot. 20: 253-259.

Butterfield, B.G. & B.A. Meylan. 1980. Three dimensional structure of wood: An ultrastructural approach. 2nd Ed. Chapman & Hall, London.

Buvat, R. 1964. Observations infrastructurales sur les parois transversales des elements de vaisseaux (métaxylème de Cucurbita pepo) avant leur perforation. Compt.-rend. hebd. séanc. Ac. Sci. Paris sér. D, 258: 6210-6212.

Carlquist, S. 1975. Ecological strategies of xylem evolution. Univ. Calif. Press, Berkeley.

Chafe, S.C. & G. Chauret. 1974. Cell wall structure in the xylem of Trembling Aspen. Protoplasma 80: 129-147.

Chattaway, M.M. 1949. The development of tyloses and secretion of gum in heartwood formation. Austr. J. Sci. Res. B 2: 227-241.

Czaninski, Y. 1977. Vessel-associated cells. IAWA Bull. 1977/3: 51-52.

Czaninski, Y. 1979. Cytochémie ultrastructurale des parois du xylème sécondaire. Biol. Cellul. 35: 97-102.

Esau, K. & I. Charvat. 1978. On vessel member differentiation in the bean (Phaseolus vulgaris L.). Ann. Bot. 42: 665-677.

Esau, K. & W.M.B. Hewitt. 1940. Structure of end wall in differentiating vessels. Hilgardia 13: 229-244.

Fengel, D. 1972. Structure and function of the membrane in softwood bordered pits. Holzforschung 26: 1-9.

Foster, R.C. 1967. Fine structure of tyloses in three species of Myrtaceae. Austr. J. Bot. 15: 25-34.

Fuji, T., H. Harada & H. Saiki. 1979. The layered structure of ray parenchyma secondary wall in the wood of 49 Japanese angiosperm species. Mokuzai Gakkaishi 25: 251-257.

Fuji, T., H. Harada & H. Saiki. 1980. The layered structure of secondary walls in axial parenchyma of the wood of 51 Japanese angiosperm species. Mokuzai Gakkaishi 26: 373-380.

Fuji, T., H. Harada & H. Saiki. 1981. Ultrastructure of the 'amorphous layer' in xylem parenchyma cell wall of angiosperm species. Mokuzai Gakkaishi 27: 149-156.

Gregory, S.C. & J.A. Petty. 1973. Valve actions of bordered pits in conifers. J. Exp. Bot. 24: 763-767.

Harada, H. & W.A. Côté. 1967. Cell wall organization in the pit border region of softwood tracheids. Holzforschung 21: 81-85.

Ishida, S. 1970. Observation of wood structure by SEM. Wood Industry 21: 560-564.

Kuster, E. 1956. Die Pflanzenzelle. G. Fischer, Jena.

Liese, W. 1965. The fine structure of bordered pits in softwoods. In: W.A. Côté (ed.), Cellular Ultrastructure of Woody Plants: 271-290. Syracuse Univ. Press, New York.

Meyer, R.W. 1967. Tyloses: development in white oak. For. Prod. J. 17: 50-56.

Meyer, R.W. & W.A. Côté. 1968. Formation of the protective layer and its role in tylosis development. Wood Sci. Technol. 2: 84-94.

Meyer, R.W. & A.F. Muhammad. 1971. Scalariform perforation plate fine structure. Wood and Fiber 3: 139-145.

Meylan, B.A. & B.G. Butterfield. 1972a. Perforation plate differentiation in Knightia excelsa R. Br. A scanning electron microscope study. Austr. J. Bot. 20: 79-86.

Meylan, B.A. & B.G. Butterfield. 1972b. Scalariform perforation plates: Observations using scanning electron microscopy. Wood and Fiber 4: 225-233.

Meylan, B.A. & B.G. Butterfield. 1981a. Perforation plate development in the vessels of hardwoods. In: J.R. Barnett (ed.), Xylem Cell Development: 96-114. Castle House Publ., Tunbridge Wells.

Meylan, B.A. & B.G. Butterfield. 1981b. Pit membrane structure in the vessel-less wood of Pseudowintera. Abstr. 13th Intern. Bot. Congr. 37.

Murmanis, L. 1975. Formation of tyloses in felled Quercus rubra L. Wood Sci. Technol. 9: 3-14.

Murmanis, L. 1976. The protective layer in xylem parenchyma cells of Quercus rubra L. Appl. Polymer Symp. 28: 1283-1292.

Murmanis, L. 1978. Breakdown of end walls in differentiating vessels of secondary xylem in Quercus rubra L. Ann. Bot. 42: 679-682.

Niedermeyer, W. 1974. Auflösung der Endwände in differenzierenden Gefässzellen. Ber. Deutsch. Bot. Ges. 86: 529-536.

O'Brien, T.P. 1970. Further observations on hydrolysis of the cell wall in the xylem. Protoplasma 69: 1-14.

O'Brien, T.P. 1974. Primary vascular tissues, In: A.W. Robards (ed.), Dynamic Aspects of Plant Ultrastructure: 414-440. McGraw Hill, New York.

O'Brien, T.P. 1981. The primary xylem. In: J.R. Barnett (ed.), Xylem Cell Development: 3-37. Castle House Publ., Tunbridge Wells.

O'Brien, T.P. & K.V. Thimann. 1967. Observations on the fine structure of the oat coleoptile. III. Correlated light and electron microscopy of the vascular tissues. Protoplasma 63: 443-478.

Ohtani, J. & S. Ishida. 1976. An observation on perforation plate differentiation in Fagus crenata Bl. using scanning electron microscopy. Res. Bull. Coll. Exp. For. Hokkaido Univ. 33: 115-126.

Ohtani, J. & S. Ishida. 1978a. An observation on the perforation plates in Japanese dicotyledonous woods using scanning electron microscopy. Res. Bull. Coll. Exp. For. Hokkaido Univ. 35: 65-116.

Ohtani, J. & S. Ishida. 1978b. Pit membrane with torus in dicotyledonous woods. Mokuzai Gakkaishi 24: 673-675.

Parham, R.A. 1973. On the substructure of scalariform perforation plates. Wood and Fiber 4: 342-346.

Petty, J.A. 1971. The aspiration of bordered pits in conifer wood. Proc. Roy. Soc. London B 181: 395-406.

Pickett-Heaps, J.D. 1967. The effects of colchicine on the ultrastructure of dividing plant cells, xylem wall deposition and distribution of cytoplasmic microtubules. Developmental Biology 15: 206-236.

Sassen, M.A.A. 1965. Breakdown of the plant cell wall during the cell-fusion process. Acta Bot. Neerl. 14: 165-196.

Schmid, R. 1965. The fine structure of pits in hardwoods. In: W.A. Côté (ed.), Cellular Ultrastructure of Woody Plants: 291-304. Syracuse Univ. Press, New York.

Schmid, R. & R.D. Machado. 1968. Pit membranes in hardwoods – fine structure and development. Protoplasma 66: 185-204.

Sheldrake, A.R. 1970. Cellulase and cell differentiation in Acer pseudoplatanus. Planta 95: 167-178.

Siau, J.F. 1971. Flow in wood. Syracuse Univ. Press, New York.

Thomas, R.J. 1967. The development and ultrastructure of the pits of two southern yellow pine species. Thesis. Duke Univ., Durham, North Carolina.

Thomas, R.J. & L.D. Bonner. 1974. The ultrastructure of intercellular passageways in vessels of yellow poplar (Liriodendron tulipifera L.). II. Scalariform perforation plates. Wood Sci. 6: 193-199.

Thomas, R.J. & K.P. Kringstad. 1971. The role of hydrogen bonding in pit aspiration. Holzforschung 25: 143-149.

Yata, S., T. Itoh & T. Kishima. 1970. Formation of perforation plates and bordered pits in differentiating vessel elements. Wood Research 50: 1-11.

The concept of cambium

PHILIP R. LARSON

North Central Forest Experiment Station, U.S. Department of Agriculture, Forest Service, Rhinelander, Wisconsin 54501, U.S.A.

Summary: The cambium is a diverse and extensive meristem, even within a single plant. Thus, cambium is not easily defined. It is best comprehended as a concept, a generalised idea of what cambium should be. As a concept, the cambium must be treated either by developmental states or in some specific way in which it is manifested. It is noted that procambium is initiated in the embryo and perpetuated in the apical shoot. The residual meristem is discussed as a transitional tissue in which the forefront of the advancing procambial strands develop. The literature on the developmental states comprising the procambium → cambium continuum are comprehensively reviewed and discussed. Metacambium is proposed as an intermediate state between procambium and cambium. Recent attempts to define the cambial state both in terms of the meristem and its derivatives are discussed.

Introduction

The cambium has been variously defined as follows: 'The actively dividing layer of cells that lies between, and gives rise to, secondary xylem and phloem (vascular cambium)' (International Association of Wood Anatomists, 1964); 'A meristem with products of periclinal divisions commonly contributed in two directions and arranged in radial files. Term preferably applied only to the two lateral meristems, the vascular cambium and cork cambium, or phellogen' (Esau, 1977); and, 'Lateral meristem in vascular plants which produces secondary xylem, secondary phloem, and parenchyma, usually in radial rows; it consists of one layer of initials and their undifferentiated derivatives' (Little and Jones, 1980).

Clearly, the cambium is a diverse and extensive meristem, and no one definition will encompass all manifestations of what anatomists consider cambium. Its diversity and extent are further exemplified by a single plant, such as a temperate-zone tree, in which procambium is initiated in the embryo and perpetuated throughout every lateral primary meristem before giving rise to cambium in the secondary body. The cambium thereafter performs its meristematic task of producing daughter cells that differentiate to specialised tissue systems. But the cambium does not remain static. Its derivatives vary either in form, or function, or rate of production at different positions on the tree, with age of the tree, and with season of the year. Moreover, the cambium responds both to internal signals and to external stimuli such as environment or wounding. Some plants possess no true cambium, others possess a cambium of limited extent or duration, while still others possess anomalous cambia. Obviously, cambium cannot be easily defined. It is best comprehended as a concept, a generalised idea of what cambium should be. As a concept, the cambium must be treated either as developmental states or in some specific way in which it is manifested.

When considered developmentally, the cambium exists as a continuum - the procambium→cambium continuum. Most present-day anatomists accept the idea of a continuum and recognise that the procambium and cambium are simply developmental states of a single meristem. Nonetheless, many contradictions and considerable controversy exist in the literature as to when and where the cambial state begins and how this state can be identified. As with all developmental continua, several transitional states in the procambium→cambium continuum can be recognised and thereby treated separately for observation, study, and discussion. In this paper, the concept of cambium will be discussed in terms of the procambium→cambium continuum and identification of the cambial state.

The discussion will be confined to dicotyledonous and coniferous species in which the cambium occurs in the normal position; anomalous cambia (Philipson and Ward, 1965) will not be considered. Also, neither the induction of cambia in cultures nor the response of cambia to experimental manipulations will be discussed.

Initiation of the procambium

The embryogeny of most dicotyledonous plants can be divided into five morphological stages (Mahlberg, 1960): 1) linear stage – a series of superimposed cells arranged in one or more rows; 2) globular stage – a symmetrical mass of cells supported at the end of the suspensor; 3) heart-shaped stage – cotyledon initiation results in a bilaterally symmetrical embryo; 4) torpedo stage – cotyledons become closely appressed to one another and hypocotyl region elongates; and 5) mature stage – condition of the embryo in ripe seed.

The procambium originates during early embryogeny, usually during the late globular stage in *Juglans* (Nast, 1941), *Phlox* (Miller and Wetmore, 1945a), *Pisum* (Reeve, 1948a), *Arceuthobium* (Cohen, 1963), *Downingia* (Kaplan, 1970) and *Stellaria* (Ramji, 1975). However, it has not been observed until the heart-shaped stage in *Quercus* (Mogensen, 1966) and *Phaseolus* (Yeung and Clutter, 1978) or as late as the torpedo stage in *Arachis* (Pillai and Raju, 1975) and *Cosmos* (Pillai *et al.*, 1975).

In *Nerium oleander* (Mahlberg, 1960), a typical dicotyledon, the procambium originates during the late globular stage in a broad zone of internal cells that develops distally from the suspensor. The procambial region is defined externally by lateral expansion of the embryonic cortex and internally by expansion of the central column of the embryonic pith. The procambium is first discernible at the level of the cotyledonary node in the region that corresponds to the future cotyledonary trace. These events occur just before the cotyledons emerge to initiate the heart-shaped stage. The apical meristems are present as less differentiated zones of meristematic cells at the suspensor and opposite ends of the globular embryo. Subsequent development of the central core of procambium is basipetal toward the incipient radicle and acropetal in the elongating cotyledons during the heart-shaped stage.

Although conifers do not exhibit a globular stage, they do progress through a similar stage of axial elongation. In *Pinus strobus* (Spurr, 1949), the procambium is first recognisable when pith tissue begins to differentiate in the axial region. Initial elongation of the procambial cells is not due to longitudinal divisions but to the fact that transverse divisions are less frequent than in cells of the adjacent pith and cortex. However, in later stages of development the procambial cells do divide longitudinally. The procambium originates slightly before or simultaneously with emergence of the cotyledons and it develops bidirectionally, extending acropetally in the cotyledons as they elongate. A similar origin of procambium has been described for *Larix* (Schopf, 1943) and *Pseudotsuga* (Allen, 1947).

A difference has been noted in the locus of procambium initiation between dicotyledons and conifers (Pillai and Raju, 1975). Whereas procambium usually originates at a relatively high level beneath the embryonic apex in dicotyledons, it originates at a lower level near the embryonic radicle in conifers. *Arachis,* a dicotyledon, corresponds to the conifer condition (Pillai and Raju, 1975).

The delineation or blocking out of procambial tissue usually occurs before the embryo has initiated growth of the cotyledons (Esau, 1965a). However, some variability has been noted. For example, procambium has been observed either slightly in advance of the cotyledonary primordia (Nast, 1941; Miller and Wetmore, 1945a; Allen, 1947; Reeve, 1948a; Mahlberg, 1960; Berlyn, 1967; Kaplan, 1970; Ramji, 1975), or simultaneously with cotyledonary primordia (Spurr, 1949; Mogensen, 1966; Mahlberg, 1960; Berlyn, 1967; Kaplan, 1970; Ramji, 1975) or following initiation of cotyledonary primordia (Pillai and Raju, 1975; Pillai *et al.*, 1975). Nonetheless, despite this variability, there is almost unanimous agreement that in all species investigated procambium develops from its site of initiation acropetally in the emerging cotyledons. Procambium within the cotyledon-hypocotyl-radicle axis therefore presents a continuum. The acropetal and continuous development of the embryonic procambium in the cotyledons has been compared to the similar development of procambium in primordial leaves of the mature shoot apex (Miller and Wetmore, 1945a). Consequently, procambium in the upper part of the hypocotyl is associated with the cotyledons as leaf traces are with leaves (Esau, 1965a). Such continuity is essential to cotyledon development, because it appears that some factor translocated from the embryonic axis by the procambium is necessary for initiating many metabolic events in the emerging cotyledons (Smith, 1974).

Much less attention has been given to procambial development of the epicotyl. Although axial polarity and a degree of apical zonation are often evident during the globular stage, the apical region from which the epicotyl will arise often remains relatively inactive until the cotyledons have emerged (Mahlberg, 1960; Pillai and Raju, 1975). For example, in *Cosmos,* the cotyledons emerged before the apex was distinguishable (Pillai *et al.*, 1975), and in *Juglans,* an apical meristem was not present until the apical notch formed between the cotyledons (Nast, 1941). Lyndon (1976) stated that the shoot meristem originates at the base of the cleft between the two cotyledons of the embryo. In *Pseudotsuga,* a shoot apex formed early but it remained inactive and contributed nothing to embryo development (Allen, 1947). The fully developed embryo of *Pinus* showed no evidence of initiating leaf primordia (Spurr, 1949), while in *Fagus* the embryos have clear procambial traces in the epicotylary apex before any leaf primordia are formed (Clowes, 1961).

Many authors have commented on the difficulties in determining when the shoot apex becomes established and in following developmental events within the apical region. It is therefore not surprising that reports of procambial development in the epicotyl are confusing and often contradictory. For example, Reeve (1948a) observed acropetal procambial development toward the epicotyl in *Pisum,* whereas Pyykkö (1974) observed basipetal development from the epicotyl toward the axis in *Honkenya.* Miller and Wetmore (1946) demonstrated that the procambial pattern established in the embryo extended into the meristem of the epicotylary shoot and leaf primordia of *Phlox.* Some authors consider them to be, at least conceptually, two independent systems (Winter, 1932; Bisalputra, 1961; Tilton and Palser, 1976), with the epicotylary vasculature being superimposed on that of the hypocotylary axis. Crooks

Table 1. Genera and families in which procambium has been observed developing acropetally and continuously

Abies	Parke, 1963 Namboodiri and Beck, 1968a	*Metasequoia*	Namboodiri & Beck, 1968b
Acacia	Boke, 1940	*Opuntia*	Boke, 1944
Actinidia	Pulawska, 1965	*Pelargonium*	Carothers, 1959
Anagallis	Vaughan, 1955	*Phlox*	Miller & Wetmore, 1945b
Arabidopsis	Vaughan, 1955	*Physocarpus*	Devadas & Beck, 1972
Araucaria	Namboodiri & Beck, 1968a Griffith, 1952	*Picea*	Namboodiri & Beck, 1968a
Capsella	Vaughan, 1955	*Pinus*	Crafts, 1943b Sacher, 1954 Namboodiri & Beck, 1968a
Cassia	Devadas & Beck, 1972	*Podocarpus*	Namboodiri & Beck, 1968a,b
Cassiope	Hara, 1975	*Potentilla*	Devadas & Beck, 1972
Cedrus	Crafts, 1943b Namboodiri & Beck, 1968a	*Prunus*	Devadas & Beck, 1972
Cephalotaxus	Namboodiri & Beck, 1968a	*Pseudotsuga*	Crafts, 1943b Namboodiri & Beck, 1968a
Chamaecyparis	Namboodiri & Beck, 1968b	*Rosaceae*	Rouffa & Gunckel, 1951
Cryptomeria	Namboodiri & Beck, 1968a	*Rubus*	Devadas & Beck, 1971
Cunninghamia	Cross, 1942 Namboodiri & Beck, 1968a	*Salix*	Reeve, 1948b Balfour & Philipson, 1962
Cupressus	Crafts, 1943b Namboodiri & Beck, 1968b	*Sambucus*	Esau, 1945
Dianthera	Sterling, 1949	*Sciadopitus*	Namboodiri & Beck, 1968a
Drimys	Gifford, 1951	*Sequoia*	Namboodiri & Beck, 1968a
Geum	Devadas & Beck, 1972	*Sequoiadendron*	Namboodiri & Beck, 1968a
Helianthus	Alexandrov & Alexandrova, 1929; Priestley & Scott, 1936 Esau, 1945	*Suaeda*	Balfour & Philipson, 1962
		Syringa	Wetmore, 1943 Vaughan, 1955
Ilex	Ashworth, 1963	*Taxodium*	Namboodiri & Beck, 1968a
Juniperus	Crafts, 1943b Namboodiri & Beck, 1968b	*Taxus*	Namboodiri & Beck, 1968a
Larix	Namboodiri & Beck, 1968a	*Torreya*	Namboodiri & Beck, 1968a
Libocedrus	Crafts, 1943b Namboodiri & Beck, 1968b	*Trifolium*	Devadas & Beck, 1972
		Tsuga	Namboodiri & Beck, 1968a
Liriodendron	Millington & Gunckel, 1950	*Vinca*	Boke, 1947
Lupinus	O'Neill, 1961	*Xanthium*	McGahan, 1955

(1933) did not detect union between the two systems in *Linum* until secondary growth was underway.

Discrepancies among the foregoing reports are due primarily to the difficulties both in detecting and in following procambial development during embryonic development and germination, and secondarily to interpretation of the observed patterns. A consensus gleaned from the published reports suggested that procambial development to the epicotyl is acropetal and continuous just as it is to the cotyledons. This pattern is perpetuated in the shoot apex during seedling ontogeny (Miller and Wetmore, 1946; Larson, 1979).

Perpetuation of the procambium

Acropetal development

It is often stated that procambium is initiated in the subapical region of the shoot. This is a misstatement. As we have just seen, the procambium is actually *initiated* in the embryo and it is *perpetuated* in the shoot during subsequent ontogeny. Perpetuation of the procambium in this manner occurs in all plants in which it develops acropetally and continuously from more mature procambium below. Such a developmental pattern has been observed almost universally in both dicotyledons and conifers (Table 1). In 54 species of the Rosaceae examined, Rouffa and Gunckel (1951) observed that not only did the procambium develop continuously and acropetally, but it was discernible up to the third or fourth peripheral, subsurface cell layers of the apex. Similarly, Devadas and Beck (1972) pointed out that in every case where developmental studies of primary vascularisation were made in both herbaceous and woody species, the procambial bundles were found to develop acropetally as independent strands. In fact, many observations have been made of precocious procambial strands advancing acropetally toward the prospective sites of their primordia before there was any evidence of these primordia on the apex (Table 2). On the basis of this anatomical evidence, many reviewers have accepted the view that procambial development in the shoot is acropetal and continuous (Esau, 1943b; 1965a; Clowes, 1961; Philipson and Balfour, 1963; Cutter, 1971; Wetmore and Steeves, 1971; Shininger, 1979).

In those plants in which the procambium has been observed to develop either basipetally from the base of a new primordium or bidirectionally from a meristem ring, it might be correct to state that the procambium was initiated in the shoot. Reports of basipetal or bidirectional development of procambium were frequent among early anatomists (see Kostytschew, 1924; Helm, 1931; Esau, 1943a; Sifton, 1944), but they have been less frequent in recent

Table 2. Genera in which precocious procambial strands have been observed*

Acer	White, 1955	*Linum*	Esau, 1942
			Girolami, 1954
Coryphantha	Boke, 1952		
		Michelia	Tucker, 1962
Dianthus	Shushan & Johnson, 1955		
		Populus	Larson, 1975
Echinocereus	Boke, 1951		
		Pseudotsuga	Sterling, 1947
Garrya	Reeve, 1942		
		Sequoia	Crafts, 1943a
Ginkgo	Gunckel & Wetmore, 1946		Sterling, 1945
Heracleum	Majumdar, 1942		

* Precocious procambial strands develop acropetally and continuously but they do so in advance
  of the primordia they will eventually serve. References cited in Table 2 do not appear in Table 1.

years (Young, 1954; Resch, 1959; Siebers, 1972; Schnettker, 1976, 1977a; Hurka and Büchele, 1976). Esau (1943b) concluded that none of the investigators whose work she reviewed produced adequate proof of basipetal procambial development. Nonetheless, some recent reviewers have subscribed to the view that procambial development is essentially basipetal (Allsopp, 1964; Wardlaw, 1965; Dormer, 1972; O'Brien, 1974; Phillips, 1976). Their conclusions were reached almost exclusively on the basis of surgical experiments on apices and/or the results of tissue and organ cultures.

Some workers postulated the *de novo* origin of new procambial strands (*e.g.,* Dormer, 1954). In keeping with his theory of meristemoids, Bünning (1965) suggested that new procambial strands develop spontaneously, with each procambial strand inducing new cell divisions in its immediate vicinity. Resch (1959) considered the 'Leptom' bundles in *Vicia* (*i.e.,* phloem bundles without evident cambium or xylem) to be initiated in this way. In a somewhat similar vein, Lang (1965) concluded that although procambial strands (his prevascular tissue) develop in an acropetal sequence, their development is not determined by acropetal forces. He suggested some form of homeogenetic induction in which a specialised tissue induces the differentiation of identical or similar tissues in its immediate vicinity (Lang, 1973). These latter views are suggestive of the idea advanced by Stebbins (1974) that the differentiation of procambium from ground tissue represents an epigenetic phenomenon. Each of the foregoing views implies bidirectional development of procambium.

It is evident that there is no universal agreement as to the direction of procambial development, although the evidence overwhelmingly favours acropetal progression. Major disagreement is between anatomists and experimental morphologists who have observed procambial patterns in intact and altered plant systems, respectively. Either the same procambial system is being interpreted differently, or the experimental procedures are altering procambial development. This discrepancy must be resolved by more critical analyses of procambial development in both intact and altered plant systems.

Although it must be recognised that plant variability permits different procambial patterns among taxa, many of the reports on basipetal development of procambium in intact plants may be due to differences in interpretation. Procambial development within the subapical region is particularly difficult to interpret because events are compressed in a narrow zone of cells. Careful analysis of thin microsections usually reveals a continuous system of acropetally developing procambial strands. The *de novo* appearance of procambial strands is most generally observed in established procambial systems. Upon closer scrutiny, many of these and similar unaccounted-for bundles will probably prove to be either equivalent to or analogous to the subsidiary bundles that differentiate basipetally from the bases of established primordia (Larson, 1975, 1982).

Residual meristem

The subapical region in which the procambium and/or procambial strands are first evident has been interpreted in different ways. In most apices, there is a ring of meristematic tissue in which procambial strands can first be detected when serial sections are followed downward from the apex. This meristematic region has been variously referred to as prodesmogen, meristem ring, procambial ring, provascular tissue, and residual meristem, and its histogenic significance has varied accordingly (Esau, 1943b). These terms were neither used nor interpreted consistently by early anatomists, primarily because of different opinions regarding where procambium originated, when it originated relative to leaf initiation, and how it was organised. For example, the idea of a meristem ring was useful to those who considered the procambium to originate within this tissue and to subsequently develop bidirectionally. On the contrary, the idea of a procambial ring was useful to those who considered the procambium to develop acropetally from more mature tissue below. These terms are still being used inconsistently today. Consequently, their intended meaning is often vague and difficult to interpret.

Introduction of the concept of residual meristem by Kaplan in 1936 (Esau, 1943b; Sifton, 1944) not only clarified procambial development but also eliminated the need for many of the descriptive terms. The residual meristem is considered a derivative of the apical meristem (Esau, 1954) or a continuation of the eumeristematic part of the apical meristem (Esau, 1965b). The distinction between apical and residual meristems is conceptual for there is no sharp dividing line between them. Nonetheless, the concept is extremely useful in describing procambial development. As noted later in this paper, the concept also provides for the differentiation of interfascicular as well as fascicular tissues.

The present concept of residual meristem incorporates within it many of the previous ideas relating to the origin of procambium. Moreover, because residual meristem applies to the status of a meristem and not necessarily to the procambium that develops within it, the concept can be acceptable both to those who adhere to bidirectional and to acropetal views of procambial development. However, most adherents to the concept favour acropetal procambial development (Esau, 1942; Sterling, 1945; Devadas and Beck, 1971). According to the latter view, a meristem ring would be analogous to the transitory region between the apical and residual meristems in which the acropetally developing procambial strands are not yet discernible; the procambial front or the region of procambial strand 'fade-out' (Larson, 1975). A procambial ring would then be analogous to the subjacent region in which the same procambial strands can first be distinguished from adjacent, uncommitted cells of the residual meristem. Provascular tissue implies that the tissue either does or will possess the potential to become vascular. Although a useful term in certain instances, provascular tissue is redundant to procambium (Esau, 1943b) and it is also redundant to residual meristem.

According to a broad interpretation, the residual meristem is not confined to the immediate subapical region but continues downward in the shoot. That is, as the procambial strands develop acropetally in the advancing front of residual meristem, the tissue surrounding the discrete bundles retains its meristematic potential. Devadas and Beck (1971) noted that the lower limit of the residual meristem is difficult to determine because additional procambial strands continue to differentiate within it at progressively lower levels. Esau (1954) also pointed out that new procambial strands serving younger leaves arise elsewhere in the residual meristem. Thompson and Heimsch (1964) concluded that a continuous procambial cylinder is formed from residual meristem as the leaf traces become confluent. Sterling (1946) and Larson (1975) refer to this tissue in which precocious procambial strands advance as 'interfascicular residual meristem'. In *Populus*, both the subsidiary trace bundles that develop basipetally from newly established leaf primordia (Larson, 1975) and the original bud traces that develop acropetally to initially vascularise axillary buds (Larson and Pizzolato, 1977), do so in the interfascicular residual meristem. The depth of the interfascicular residual meristem is not great, because these latter events all occur in association with the closely spaced primordia immediately below the apex.

Sterling (1946) described procambial development in the residual meristem of *Sequoia* shoots in extreme detail. The first indication of the presence of a procambial strand was the deeper staining of a small group of irregularly arranged cells in the residual meristem. As the strand developed further, a decrease in transverse diameter of these cells became evident. Somewhat later, cells of the cortical parenchyma vacuolated and differentiated to establish a continuous parenchymatous sheath. The cylinder of procambial strands was thus delimited externally by the cortex and internally by the parenchymatous pith that differentiated earlier. The procambial cylinder consisted of acropetally developing procambial strands in different states of development interposed with interfascicular residual meristem.

Sterling's (1946) description of early procambialisation in the residual meristem of *Sequoia* is typical of that found in most dicotyledonous and coniferous plants. A review of many reports on apical development suggests a similarity in pattern but variability in minor details. There is no sharp dividing line between the apical and residual meristems. Rather, there exists a transitional continuum in which cells of the residual meristem gradually assume characteristics different from those of the apical meristem. These changes are accompanied by the gradual parenchymatisation of the incipient cortex and pith, which contribute to the primary elongating meristem (Sachs, 1965; Loy, 1977). One effect of the acropetally developing procambial strands is to block out pattern (Wetmore *et al.*, 1964) in the residual meristem by separating the fundamental tissues of cortex and pith (Wetmore and Steeves, 1971). Interestingly, cell division and elongation are induced across the entire subapical region including cortex, pith, and vascular tissue, but the distribution of these events varies in the different tissues (Sachs, 1965). Thus, the initial elongation of procambial cells is primarily by passive stretching; *i.e.*, by accommodation to the relatively greater activity of the pith and/or cortical cells.

Differentiation of the fundamental tissues of cortex and pith are generally assumed to occur basipetally, whereas the procambial strands that separate them develop acropetally. But to what extent are these tissues developmentally interdependent? Is it possible that the acropetally developing procambial strands in some way regulate differentiation of the pith and cortical tissues in addition to blocking out the position of the future vascular system? In order to develop acropetally, the procambial strands must receive growth requirements from antecedent leaves below (Larson, 1982). The procambial strands are ideally situated to provide growth requirements to the incipient cortex and pith as well as to the procambial front in the residual meristem. Therefore, it is conceivable that the advancing procambial strands block out pattern both directly at the procambial front and indirectly in the fundamental tissues of cortex and pith which they segregate.

Do procambial strands advance?

Clowes (1961) criticised use of the term 'acropetal' because it implies a progression toward the apex. He suggested that as the apex grows away from the base of the plant the procambium simply follows it, maintaining a fixed distance below the apex. Priestley and Scott (1936) also stated that procambial strands keep pace with elongation of the tissue in which they are situated. Allsopp (1964) considered this process to be a natural consequence of apical growth. However, evidence indicates that the term acropetal is perhaps correctly used with regard to procambial strands. It is tacitly assumed in the writings of all investigators who have observed acropetal development of procambial strands that their development is progressive. Such an assumption is borne out by the many reports that procambial strands at different stages of development occur in the residual meristem (Esau, 1942; Sterling, 1946, 1949; O'Neill, 1961; Balfour and Philipson, 1962; Parke, 1963; Devadas and Beck, 1971; Larson, 1975). Differential development is an essential requirement for precocious procambial strands.

An argument suggesting that procambial strands not only advance in the residual meristem but that they do so at different rates is presented graphically in Fig. 1. If it is assumed that procambial strands maintain a constant rate of advance and a uniform distance beneath the apex, the required number of precocious procambial strands would be equivalent to the number of vascular sympodia; e.g., 5 procambial strands in a 2/5 phyllotactic system and 13 in a 5/13 phyllotactic system (see also Crafts, 1943a). On the contrary, if it is assumed that procambial strands can advance at variable rates, fewer procambial strands are required depending on their time of divergence and rate of advance. The latter alternative is closer to the observed facts; the number of precocious procambial strands is always much less than the number of sympodia. The data in Figure 4 of Larson (1975) indicate that precocious procambial strands do indeed advance in the residual meristem at different rates, and Crafts (1943a) commented on the rate at which procambium appeared to develop acropetally in *Sequoia*. Warren-Wilson and Warren-Wilson (1981) suggested that the gradient-induction hypothesis may account for the advance of procambium, whereas Ans (1979) preferred a reaction-diffusion mechanism.

If the existence of precocious procambial strands is denied, other problems arise (Larson, 1982). Either procambial strands must develop basipetally to

96

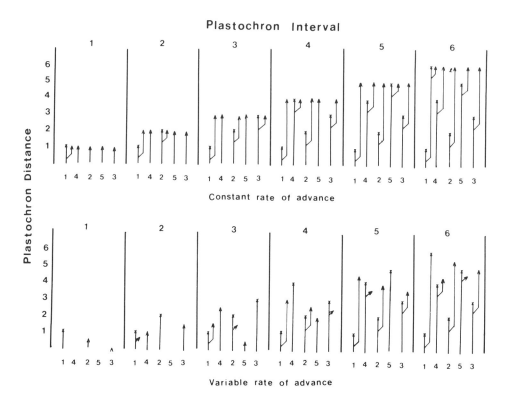

Fig. 1. Acropetal progression of procambial strands in residual meristem for constant (top) and variable (bottom) rates of advance. Basic assumption is a 2/5 phyllotaxy with one central leaf trace terminating in a primordium (X) and simultaneously giving rise to a new procambial central strand (↑) during each plastochron. Sympodia are numbered on the abscissa. Plastochronic distance is the spatial distance traversed by both the apex and the procambial strands during one plastochron. Constant rate of advance: New procambial strands must advance at same rate as apex; therefore, the number of precocious procambial strands (5) equals the number of vascular sympodia. Variable rate of advance: Assuming that procambial strands advance at a rate of 1½ plastochronic distances/plastochronic interval, only 4 precocious procambial strands are required to maintain a steady state. By assuming either a greater rate of advance or a variable rate of advance for procambial strands at different states of development, fewer precocious strands would be required.

unite with previously established leaf traces, or they must develop acropetally through many internodes during each plastochron. Both alternatives are unrealistic because they would require rates of either ascent or descent far greater than those required by precocious procambial strands that develop acropetally at variable rates.

Are procambial cells differentiated?

Shininger (1979) questioned whether procambium should be considered a differentiated tissue. He noted that until overt cytodifferentiation occurs, procambial cells cannot be defined in terms of either their developmental commitment or their real developmental state. This view reflects Newman's (1961) earlier contention that it is the surrounding tissues that differentiate to parenchyma; the procambium remains meristem.

The term 'differentiation' can have several connotations; Shininger (1979) adopted one of the more restrictive ones. However, in general biological usage differentiation connotes both states and processes; cells, tissues, and organs differentiate when they pass from one state to another (Heslop-Harrison, 1967). With regard to meristematic tissues, differentiation can therefore refer to the potential of those tissues to produce new meristematic states. In this sense, derivation of residual meristem from apical meristem and procambium from residual meristem would be analogous to the earlier derivation of an apical meristem from embryonic tissues as discussed by Wetmore and Steeves (1971). Each of these meristematic states conveys a new potential for differentiation to its derivative tissues. Meristematic differentiation, according to this view, becomes inextricably bound with meristematic determination (Wareing, 1978).

Many investigators have commented on the difficulty in identifying the first procambial cells (McGahan, 1955). In fact, it is not possible to recognise single, isolated procambial cells. Only when a group or island of cells develops can a procambial strand be recognised. Moreover, at the present time it is not possible to identify presumptive procambial cells either cytologically or biochemically, a fact cited by Shininger (1979) to support his interpretation of differentiation. Nonetheless, as procambial strands develop the constituent cells differ considerably from those of either the apical or residual meristems, and late procambium differs considerably from early procambium. Thus, the procambium retains its meristematic potential even though it progresses through a continuum of differentiation states.

Procambium → cambium continuum

Procambial state

As noted earlier, there is no sharp dividing line between the apical and residual meristems. Rather, there exists a transitional continuum in which cells of the residual meristem gradually assume characteristics different from those of the apical meristem. These changes are accompanied by gradual parenchymatisation of the prospective pith and cortex, which contribute to the primary elongating meristem (Sachs, 1965; Loy, 1977). One effect of the acropetally developing procambial strands is to block out pattern (Wetmore *et al.*, 1964) in the residual meristem by separating the fundamental tissues of pith and cortex (Wetmore and Steeves, 1971).

In *Linum*, Esau (1942) was able to follow procambium 'to the highest level of the apex', to within a few cells of the youngest leaf primordium. These procambial cells were small in transverse diameter, even smaller than those of the apex, and short in length. The narrow, elongated shape typical of procambial cells was acquired gradually at successively lower levels in the meristem. The procambial cylinder was delimited first by vacuolation of pith cells followed somewhat later by those of the cortex. At the level in the meristem where the procambial cylinder became readily recognisable, it consisted of discrete procambial strands alternating with vacuolated interfascicular areas. As a rule, cells of the interfascicular areas neither vacuolated nor enlarged to the same extent as those of the pith and cortex.

Esau's (1942) description of events at the forefront of developing procambium is typical of that found in both dicotyledonous and coniferous plants. In general, procambial cells are distinguished from adjacent cells of the residual meristem by their narrow, elongated shapes, denser cytoplasmic contents and/or stainabilities, and arrangement in groups or bundles. However, the typical elongated shape of procambial cells is not acquired by a relatively greater rate of cell elongation but rather by passive accommodative growth. One of the first stages of elongation that consistently occurs in all plants is a rapid increase in cell number (Sachs, 1965). This increase is primarily by anticlinal divisions and it occurs uniformly across the entire subapical region including prospective pith, cortex, and vascular tissues. Anticlinal divisions usually precede periclinal divisions in the procambial strands and undoubtedly account for the short cells immediately adjacent to the youngest primordia observed by Esau (1942) in *Linum*. The procambial cells do not cease dividing anticlinally. While a procambial strand or trace enlarges in diameter by periclinal divisions, it also increases in length by anticlinal divisions that keep pace with the accelerated rate of internodal elongation (Catesson, 1974).

Cells of the prospective pith, or rib meristem, usually vacuolate and enlarge before those of the cortex in both dicotyledons and conifers; rib meristem cells often vacuolate especially early in some conifers. Nevertheless, the time at which vacuolation begins and the rate at which it proceeds vary among species. Once vacuolation is underway in both tissues, the rate of cell elongation relative to cell division is most often greater in cortical than pith tissues

Table 3. Increase in length of procambial cells during growth of the internode in *Acer*

Internode length	0	100 μm	1 cm	5 cm	15 cm
Procambial cell length	10 ↓ 12 μm	15 ↓ 20 μm	50 ↓ 70 μm	200 μm	250 μm

From Catesson, 1974 (with permission of McGraw-Hill).

(Sachs, 1965). In any event, the procambial cells elongate by a passive process in which their growth is accommodated to that of the vacuolating and expanding pith and cortical cells (Esau, 1942; Gunckel and Wetmore, 1946; Newman, 1961; Larson, 1975). By this process, procambial cells increase considerably in length during internodal elongation (Table 3).

Procambial strands enlarge by cell divisions within the strand and by the acquisition of cells from adjacent tissues. Cell acquisition often precedes periclinal cell division, and the earliest stages of strand enlargement are by this process. The first acquired cells should be properly considered derivatives of the residual meristem (Esau, 1954; Parke, 1963; Larson, 1975); Clowes (1961) urged caution in use of the term residual meristem because cells may be later added from the fundamental tissues of both pith and cortex. Later-acquired cells divide periclinally and the daughter cells enlarge the periphery of the procambial strand. Such additions appear more frequent on the cortical side, thus enlarging the prospective phloem region of the strand. Periclinal divisions also occur throughout the interior of the strand. These divisions are at first irregular with no preferred direction of orientation, although the division plane may parallel the strand periphery. As in the case of the earlier cell divisions, the daughter cells of periclinal divisions elongate by passive accommodative growth.

Although procambial strands develop acropetally, procambium has been described as if it developed basipetally. Protophloem and protoxylem differentiation occur at various times during early procambialisation depending on the ontogenetic stage of development of the plant and the species investigated. These differentiation events will not be discussed in this paper other than to point out that the evidence overwhelmingly supports the acropetal differentiation of protophloem in continuity with more mature protophloem below. Protoxylem, on the contrary, differentiates bidirectionally. It usually, but not always, is initiated in the vicinity of the primordium base, and differentiation proceeds basipetally in the leaf trace and acropetally in the node. Protophloem differentiation almost always precedes that of protoxylem.

The differentiation of protophloem and protoxylem provides a more definitive functional role for the procambial system. However, these tissues comprise just one stage of the overall continuum of vascular development and the manner in which they differentiate must be kept in perspective. Just as the procambium blocked out in the residual meristem consists of procambial strands at different stages of development, the procambial cylinder at the levels of both protophloem and protoxylem differentiation also consists of procambial traces and strands at different stages of development (Larson, 1980a). All procambial traces and strands are larger and more completely differentiated at their levels of divergence from a parent trace than at successively higher levels in the shoot because of their acropetal development. Even precocious procambial strands that diverge in advance of the appearance of their prospective primordia have protophloem maturing acropetally within (Sterling, 1946; Larson, 1976a). This differential development of procambial traces at any one stem level results from the fact that the traces not only serve primordia at different stages of development, but they also diverge from parent traces serving leaves at different stages of development (Larson, 1976b). And so, although each leaf trace will eventually progress through the same ontogenetic sequence as every other leaf trace, it will not do so at the same level in the stem. Failure to appreciate this developmental process has undoubtedly contributed to the many reports of procambial bundles, phloem-only bundles, accessory bundles, *etc.* that have been observed distributed among the more mature trace bundles in the procambial cylinder.

Metacambial state

Esau (1943b, 1954, 1965a) has repeatedly emphasised that a distinction must be made between primary and secondary growth. Primary growth begins in the embryo and continues until internodal elongation ceases. During this time the lateral meristem remains procambial. Nonetheless, vast changes occur both in the procambium and its derivatives during internodal elongation, and these changes have contributed to many contradictions and much confusion regarding the nature and extent of this meristem, Anatomists have recognised and acknowledged changes in the procambial derivatives by distinguishing metaphloem and metaxylem from protophloem and protoxylem, although not necessarily by universally accepted definitions. However, equivalent stages have been neither recognised nor proposed for the lateral meristem.

Larson (1976a) attempted to resolve this dilemma in *Populus* by sub-dividing the procambium into procambium and metacambium. Recognition of a metacambium does not contravene the original concept of a procambium → cambium continuum. It simply recognises an essential transitional stage – the procambium → metacambium → cambium continuum. Metacambium is distinguished from procambium both by the meristematic cell division planes and by the products of these divisions.

The procambium consists of irregularly oriented cells (Fig. 2), and cell division planes are also irregularly, but not necessarily randomly, oriented. Protophloem and protoxylem differentiate from derivatives of the phloic and xylary procambium, respectively (Esau, 1953). The metacambium is preceded by the initiating layer, which first occurs as isolated, periclinal divisions in the incipient metacambial region (Fig. 3). As the periclinal divisions increase in frequency, cells of the initiating layer occur as discontinuous bands within adjacent bundles (Fig. 4) and eventually as a continuous band extending across contiguous bundles and leaf traces to form the metacambial cylinder (Fig. 5).

The metacambium, preceded by the initiating layer, advances acropetally and continuously in each procambial bundle. It advances first in the original procambial bundle of each leaf trace, the one that developed acropetally, and then in the subsidiary bundles that differentiated basipetally. Thus, the metacambium advances not only acropetally but also laterally within each leaf trace. Moreover, it advances acropetally and helically within the procambial cylinder in conformance with the age and position of each developing leaf in the helical phyllotaxy.

Metacambium is also defined on the basis of its derivatives. For example, in *Populus*, both protoxylem and metaxylem are initiated in proximity to the primordium base and differentiation proceeds bidirectionally. Protoxylem originates deep within the trace, well separated from the region in which the metacambium will later arise. Although they differentiate from irregularly oriented procambial cells, protoxylem elements may occasionally occur in short, disjunctive, radial files. By contrast, metaxylem elements are derivatives of metacambial divisions and they therefore occur in consecutive files of radially arranged cells. Some metaxylem elements are derivatives of early divisions occurring in the initiating layer. Although usually well separated at their level of origin near the primordium base, these early metaxylem elements often appear radially contiguous to protoxylem elements at lower levels in the trace. Obviously, the distinctions between procambium and metacambium and their derivatives are not precise. As with every attempt to subdivide a continuum, certain arbitrary decisions and definitions must necessarily be employed.

Subsequent development of the metacambium results in a well-developed

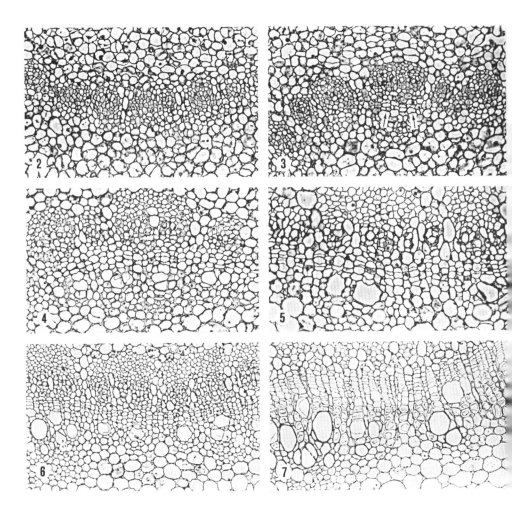

Fig. 2-7. Development of the procambium→metacambium→cambium continuum proceeding basipetally within the elongating terminal shoot of a *Populus deltoides* plant. − Fig. 2. Procambium: irregular cell divisions with no tangential continuity. − Fig. 3. Early initiating layer: isolated bands of tangentially oriented periclinal divisions (arrows) usually initiated in the original bundle of a central leaf trace and then in subsidiary bundles. − Fig. 4. Late initiating layer: tangentially oriented divisions almost continuous across bundles of a central leaf trace but with many discontinuities. − Fig. 5. Early metacambium: tangentially oriented divisions essentially continuous around entire vascular cylinder. − Fig. 6. Mid-metacambium: active division of metacambium and increase in differentiation of its derivatives. − Fig. 7. Late metacambium: very active division of metacambium and very active differentiation of its derivatives; just prior to cambial state. − Fig. 2-5, ×200. Fig. 6-7, ×130.

lateral meristem (Fig. 6). In the final stages of internodal elongation, the metacambium is anatomically indistinguishable from the cambium (Fig. 7). However, its derivatives are metaxylem and metaphloem and it must therefore be regarded as part of the primary plant body. The transition from metacambium to cambium will be discussed in the next section.

Cambial state

Many criteria have been used in attempts to distinguish cambium from procambium. The level within the shoot at which these characteristic features occur often varies among species and it may vary within a plant during ontogeny. Moreover, a characteristic feature may either occur in one species and not in another or be of little significance in another species. Even when a characteristic feature is present, one investigator may accord it greater significance or interpret it differently than another investigator. Difficulties are most often encountered because the procambium and cambium exist as a continuum and there is no distinct boundary between them. When followed downward in the vasculature, each feature or event changes gradually and a change in one event may overlap that of another. Consequently, it is difficult to evaluate the literature on this subject because of the differences that occur among species and in the interpretation of observed events.

Esau (1954) has repeatedly stressed the distinction between primary and secondary growth of the plant body. Briefly, the primary body would include all vascular tissues formed during internodal elongation; this would include both the procambial and metacambial stages previously discussed. The secondary body would then include all vascular tissues formed following the cessation of internodal elongation. However, even this definition is difficult to apply because of the continuum of events that occur in the vascular system.

Events occurring during internodal elongation should logically be discussed in the previous metacambial stage. However, the idea of a metacambium has been just recently introduced. It was therefore decided to discuss these events not only in terms of procambium *vs.* cambium as the original authors intended, but also by considering whether the authors recognised these events as occurring either before or after internodal elongation.

Before internodal growth cessation. – Radial seriation of cells and derivatives of the lateral meristem was the characteristic most commonly associated with cambium by early anatomists. However, as Esau (1943b) has pointed out, radial seriation alone is not a valid criterion for distinguishing cambium from procambium. Radial seriation can occur extremely early in procambial development, often coinciding with the beginning of rapid cell expansion and internodal elongation. The vascular elements of some conifers have been observed to be radially aligned particularly early in the procambial stage (Sterling, 1946; Sacher, 1954; Parke, 1963; Soh, 1972). Sterling (1947) noted that radial seriation was present in the procambium before either measurable shoot elongation or vascular differentiation began in *Pseudotsuga*. He regarded radial seriation as an advanced stage of procambial development. Yet, Gunckel and Wetmore (1946) did not detect radial seriation until the procambium was well established in *Sequoia*.

Radial seriation during procambial development is also common among dicotyledons and its occurence has been documented in many species (Esau, 1943b). However, the procambial derivatives of dicotyledons appear to be more irregulary arranged than those of conifers and radial seriation is considered to occur later during procambial development. This latter discrepancy may be more apparent than real and it may simply represent different interpretations of when radial seriation actually begins.

Elements of the protoxylem are more likely to occur in radial rows than those of the protophloem, although Sterling (1946) and Fahn *et al.*(1972) noted that radial seriation of the protophloem preceded that of the protoxylem in *Sequoia* and *Ricinus*, respectively. The radially aligned protoxylem and/or protophloem elements may differentiate either directly from contiguous procambial cells or from daughter cells of recent periclinal divisions within a procambial strand. In some instances, elements initiated in radial rows may later be shunted into irregular arrangements by the expansion and accommodation of other cells in the procambial strand. Like most protoxylem and protophloem cells, the first radially aligned elements are usually stretched and obliterated during subsequent internodal elongation. In some species, the protophloem elements retain their protoplasm and nuclei up to the time of obliteration (Sterling, 1947; Parke, 1963).

Despite its rather common occurrence among many plant species, radial seriation of protoxylem and protophloem elements cannot be considered either as a criterion of or even as an indicator of impending cambial development. Not only do these events occur early during primary stem elongation, but they originate from cells that can in no way be considered a tangentially oriented meristem.

With increased development of the procambium, or as one progresses downward in analysing the procambium, the width of the radially seriated region in each procambial trace increases tangentially. In the transverse plane, the periclinal divisions appear to spread laterally in the region often referred to as the wings of the procambial trace (Gunckel and Wetmore, 1946; Sterling,

1946; Thompson and Heimsch, 1964). The wings are formed earlier by the acquisition of cells from residual meristem. The periclinal divisions result in a more or less continuous, tangentially oriented band of procambial initials, but still confined within the trace.

When viewed in the tangential plane at the aforementioned stage of development, the recently divided procambial cells may appear storied. The storied arrangement has been observed in both dicotyledons (Fahn *et al.*, 1972) and conifers (Sterling, 1947; Sacher, 1954; Parke, 1963), but it is particularly conspicuous in *Pseudotsuga* (Sterling, 1947); Soh (1972) found no evidence of a storied procambium in *Ginkgo*. Storied procambial cells are usually described as having gabled end walls. Both the storied arrangement and the shape of procambial cells at this stage of development are undoubtedly due to accommodative growth following cell division as described earlier. However, no consistent explanation can be offered as to why some species display a storied procambium and others do not. The presence of storied procambium bears no consistent relation to a storied stratified cambium. Neither *Robinia* (Soh, 1947b) nor *Hoheria* (Butterfield, 1976), species with storied cambia, exhibited storied procambium.

The stages at which the foregoing procambial events occur correspond to the initiating layer and early metacambium described by Larson (1976a). The first periclinal divisions would be indicative of the initiating layer, and the integration of these divisions in a tangentially continuous band within a procambial trace or within bundles of a trace would conform to early metacambium. Although no unanimity would be found in the literature as to the designation of derivatives of the transitional initiating layer, most investigators would probably agree that derivatives of early metacambium differentiate to metaxylem and metaphloem. These events occur during rapid internodal elongation, consequently during primary development. Most present-day investigators would consider this stage procambium.

During subsequent development the procambium, or the metacambium if one accepts this term, assumes characteristics more closely resembling those of cambium. The width of the zone of periclinally dividing cells increases radially in the main body of each procambial trace and the zone spreads laterally in the interfascicular regions to eventually form a continuous tangentially oriented cylinder. Although still procambial, the cylinder possesses both the general appearance and many of the structural characteristics attributed to cambium when viewed in the transverse plane. When viewed in the longitudinal plane, the procambial cells may appear elongated with tapering or pointed end walls suggestive of intrusive growth. Measurements made by Catesson (1974) in *Acer* indicate that cells of the procambium increased in length manyfold during elongation growth of the internode (Table 3).

Many investigators have described the foregoing developmental events according to arbitrary stages and it is therefore difficult to relate them to uniform morphological benchmarks. In this discussion, is has been assumed that they occur during primary growth, consequently during internodal elongation. Priestley and Scott (1936) adopted an extreme view and considered al radially seriated derivatives as secondary, whereas Amer and Neville (1979) simply stated that cambial activity began while internodal elongation was still underway. Sterling (1946) identified four criteria for distinguishing cambial from procambial cells: 1) when they begin intrusive growth; 2) when the radial walls thicken more than the tangential walls; 3) when ray initials are distinct from neighbouring fusiform initials; and 4) when vascular elements (phloem) begin to differentiate from cells formed by successive periclinal divisions. He concluded that, except for the formation of a closed cylinder of cambial initials, the future mode of cambial activity has been fully established. Although these criteria are normally associated with secondary growth, Sterling (1946) found that they occured during the period of rapid internodal elongation in *Sequoia*.

After internodal growth cessation. – According to Eames and McDaniels (1925), the cessation of stem elongation is either accompanied by or immediately followed by cambial development in those plants exhibiting secondary growth. Most anatomists would now agree with this conclusion. However, surprisingly little research has been undertaken to describe the primary-secondary transition region and the events that define it. The most obvious events are changes in the shape and growth pattern of fusiform initials, the initiation of secondary rays, and the differentiation of secondary vascular elements. Each of these events has been associated either singly or in combination with the meristematic state designated as cambium.

As in the case of radial seriation discussed previously, each event associated with the primary-secondary transition occurs gradually and these events do not necessarily occur in unison. Catesson (1964) observed that the transition in *Acer* was abrupt and many reviewers have commented on her observation. However, abruptness is again a matter both of degree and interpretation. For example, Larson and Isebrands (1974) noted that although the transition occurred gradually in what they referred to as the 'transition zone' of *Populus*, a specific level in the transition internode could nonetheless be identified as the 'transition plane'. Every attempt to subdivide a continuum is both relative and arbitrary, and this fact most certainly becomes evident in defining the cambium.

One of the most characteristic features of the vascular cambium is the presence of two types of meristematic cells, the elongated fusiform initials and the more or less procumbent ray initials (Philipson and Ward, 1965). However, the procambium preceding this stage appears as a relatively homogeneous tissue in tangential view (Esau, 1943b). Soh (1972, 1974a, 1974b) commented on the homogeneous structure of the procambium in all the species he examined as did Enright and Cumbie (1973). Segregation of the procambium into two tissue systems occurs gradually and it apparently does so at earlier stages of procambial development in some species than in others. Cumbie (1967) noted that anticlinal divisions leading to the formation of a system of short cells appeared particularly early during procambial development in *Canavalia*. Some workers interpreted the appearance of two distinct tissue systems as contributory evidence of cambium, even though it occurred in internodes that were still elongating (Sterling, 1946; Philipson and Ward, 1965). Other workers considered the appearance of rays as just one manifestation of the procambial-cambial transition preceding the cessation of internodal elongation (Thompson and Heimsch, 1964; Cumbie 1967; Butterfield, 1976), whereas still others considered it one of the decisive events in cambial development (Catesson, 1964). These contradictory interpretations are undoubtedly due not only to differences among species in the relative time at which the two tissue systems appear to segregate, but also to the extremely gradual transition that occurs in all species. Fahn *et al.* (1972) concluded that it was impossible to determine either the time or position at which rays first appeared in *Ricinus*. These criteria do not of course apply to rayless woods.

Segregation of the lateral meristem into two tissue systems requires the differentiation of two quite distinctive types of meristematic cells, the fusiform and ray initials. As noted earlier, procambial cells at first exhibit little or no intrusive growth, their end walls may appear gabled or only slightly tapered, and they sometimes occur in a storied arrangement. As internodal growth accelerates and the procambial zone in each leaf trace increases both radially and tangentially, the procambial cells become more elongated and their end walls more tapered. Intrusive tip growth is common during rapid internodal elongation, but its time of inception varies widely among species (Thompson and Heimsch, 1964; Cumbie 1963, 1967; Soh 1972, 1974a, 1974b; Enright and Cumbie, 1973).

Intrusive growth of the procambial cells begins before the appearance of rays, although the procambium may appear less homogeneous than it did earlier. Differentiation of rays occurs in two ways. They are either derived from certain primordial cells of the interfascicular segments that did not undergo extensive vertical elongation or from elongated procambial cells of the fascicular segments that divided anticlinally (Barghoorn, 1940a). It has been suggested that the procambial cells, either fascicular or interfascicular, that will later give rise to cambial ray initials be referred to as 'primordial ray initials' (Barghoorn, 1940b) and those that will give rise to cambial fusiform initials be referred to as 'primordial fusiform initials' (Cumbie, 1967).

Both intrusive growth of the elongated procambial cells (the incipient fusiform initials) and blocking out of the prospective ray tissue occur during the final stages of internodal elongation, during the late metacambial stage of Larson (1976a). Nonetheless, neither the meristematic tissues nor their derivatives assume the true characteristics attributed to cambium until internodal growth has terminated. They are therefore part of the primary body and they should be considered as transitional or incipient stages of cambium.

It has been repeatedly stressed that events leading to the identification of cambium occur gradually. This is a consequence of the constant accommodation of procambial cells and their derivatives to internodal elongation. The lateral meristem cannot stabilise until internodal elongation has ceased. Several investigators described events that occurred during the final stages of internodal elongation and attempted to define cambium in terms of these events (Catesson, 1964; Thompson and Heimsch, 1964; Cumbie, 1967; Enright and Cumbie, 1973; Butterfield, 1976). Fahn *et al.* (1972) also described these events in detail but they did not relate them to internodal elongation. None of the attempts to distinguish cambium from procambium was successful when based exclusively on either the appearance or structure of the meristematic cells.

The most definitive criterion of cambium is the nature and structure of its derivatives. Much emphasis has been placed on the fact that protoxylem is capable of stretching and becomes obliterated whereas metaxylem is incapable of stretching and remains functional although both elements differentiate during internodal elongation (Frey-Wyssling, 1940; Eames and MacDaniels, 1947). However, it is possible that some early metaxylem elements are stretched and rendered non-functional and that the first secondary xylem elements are initiated during the final stages of internodal elongation.

Bailey (1944) observed an abrupt decrease in the length of tracheary elements during the transition from primary to secondary xylem and suggested that this decrease might be used to distinguish the two growth regions. Cumbie (1963) confirmed this abrupt decrease in *Hibiscus,* but not in *Canavalia* (Cumbie, 1967). In *Canavalia,* Cumbie (1967) found that although the first-formed secondary xylem elements were shorter than the last-formed primary elements, the most abrupt decrease in length occurred in the primary xylem. This decrease was the result of numerous anticlinal divisions of the procambial cells. In some herbaceous species there is no evidence of an abrupt decrease in xylem element length in the transitional region of primary-secondary growth (Carlquist, 1962).

Bailey (1944) also observed that whereas metaxylem vessels elongate considerably during differentiation, secondary xylem vessels do not; they remain essentially the same length as the cambial initials from which they originate. He suggested that this relation might be used as an indicator of cambium. Isebrands and Larson (1973) demonstrated that both the secondary vessels and sieve elements maintained this relationship in *Populus*. Although this criterion might be a valid test of cambium in some species, it is difficult to detect and requires precise radial sectioning.

At the termination of internodal growth, tissues of the primary body give way to those of the secondary body. Considerable overlap occurs during this transition, particularly in the last internode to mature. Nonetheless, tissues of the secondary body maintain both developmental and functional continuity with those of the primary body. For example, metaxylem is not 'transformed' to secondary xylem. Rather, metaxylem vessels anastomose with secondary xylem vessels in the transition zone and functional continuity is maintained (Larson, 1980a). Consequently, while the metaxylem vessels are maturing, secondary xylem vessels are differentiating.

Larson and Isebrands (1974) used these events to identify the primary-secondary transition in *Populus*. Although secondary elements can be observed differentiating during the final stages of internodal elongation, they do not mature until internodal elongation has ceased. At this time the elements attain their final lengths and their walls begin to lignify. Lignified elements can be readily detected by their birefringent walls in polarised light. Secondary vessels are sometimes difficult to distinguish from late metaxylem vessels in transverse sections, but xylem fibres are easily identified. Because fibres are associated with secondary vessels but not with metaxylem vessels, the primary-secondary transition was judged to occur '...when fibres with birefringent walls were first detected both within and between adjacent traces forming the vascular cylinder, with the exception of those traces last to enter the stem and all traces situated between them'. (Fig. 8 & 9). The foregoing definition applies to the 'transition plane', the position in the stem that satisfies the specific requirements at the time of sampling. However, differentiation events occur gradually and the broader region of the internode in which differentiation is proceeding is referred to as the 'transition zone'. The transition zone proceeds acropetally and helically in the maturing internode in consonance with the maturation of leaf traces in the genetic helix.

The foregoing definition applies to derivatives of the cambium and not to

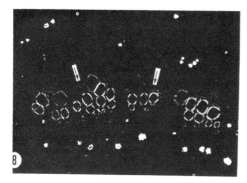

Fig. 8-9. Polarised light micrographs of stem transections of a *Populus deltoides* plant in the vicinity of the primary-secondary vascular transition zone (TZ). – Fig. 8. Approximate level of TZ. Metaxylem vessels are highly birefringent. Weakly birefringent fibres (arrows), indicative of the cambial state, are differentiating external to the metaxylem vessels and between the differentiating secondary xylem vessels. Although phloem bundle-cap fibres are differentiating, their walls do not exhibit birefringence. – Fig. 9. Immediately below the TZ, well developed secondary xylem consisting of differentiating fibres, vessels, and rays. Note the well-developed phloem bundle-cap fibres. The strongly birefringent cells are crystalliferous, × 75.

the cambium *per se*. Although arbitrary, it has been found to be remarkably consistent and reproducible as well as simple to use and evaluate. It has been used to evaluate the primary-secondary transition in *Populus* plants at different ontogenetic stages from seedlings (Larson, 1979) to relatively large-sized plants (Larson, 1980b), and from plants progressing into dormancy (Goffinet and Larson, 1981; Larson and Goffinet, 1981) to the induction of secondary xylem in lateral branches (Richards and Larson, 1982).

Larson and Isebrands (1974) relied primarily on the presence of fibres and vessels as criteria of secondary development because of their consistent appearance and ease of detection. However, other elements and tissues also attain secondary characteristics at the termination of internodal elongation. The rays take on secondary characteristics and their cells lignify, often progressing inward to the late metaxylem (Thompson and Heimsch, 1964). Catesson (1964) considered the differentiation of secondary rays to be the main distinction between primary and secondary development in *Acer*. Groups of sclerenchymatous cells often surround the obliterated protoxylem (Esau and Morrow, 1974; Larson, 1976b) and primary phloem fibres, or bundle-cap fibres, differentiate contiguous to the obliterated protophloem (Blyth, 1958; Esau and Morrow, 1974; Larson and Isebrands, 1974). Bond (1942) found that the lignification of pericycle coincided with the concluding phase of primary growth

Table 4. The major differences between a young procambium and the mature cambium of *Ricinus*

	Procambium	Cambium
1.	Cells with gabled endings in radial view	Cells with flat endings in radial view
2.	Cells relatively short	Cells long
3.	Storied arrangement	Non-storied arrangement
4.	The protoplast stains deeply	The protoplast does not stain in a special way
5.	There are no ray initials	Vascular ray initials present
6.	Planes of division of neighbouring cells, as seen in cross-sections, do not correspond	Planes of division of neighbouring cells do correspond

From Fahn *et al.*, 1972 (with permission of A.K.M. Ghouse).

while the initiation of phellogen marked the beginning of secondary growth in *Camellia*. Schnettker (1977b) considered the first evidence of cambial activity and the onset of secondary growth to occur simultaneously in *Helianthus*. Although these and other events can be correlated with the cessation of internodal elongation, differentiation occurs gradually within the maturing internode, the events do not necessarily coincide with one another, and most of them are difficult to evaluate as reproducible criteria.

The consensus of most investigators has been that the procambium and cambium exist as a continuum. Beginning with the earliest detectable procambium in the residual meristem or beneath the apex, the procambium develops imperceptibly through a series of stages. As in every other continuum, differences can be noted when discrete parts of this continuum are compared, but the transitional regions between the parts nonetheless present a continuum. This relation also holds when the transitional region between procambium and cambium is examined. Thoday (1922) concluded that primary and secondary xylem were produced by the same cambium and that they could only be distinguished histologically. Catesson (1964), however, argued that cambium was distinct from procambium and that it represented a new meristematic state. Although Fahn *et al.* (1972) could find no consistent characteristic distinctive of cambium, they did compile a list of events that differed in degree between procambium and cambium (Table 4). Because of the gradual changes that occur within the procambium → cambium continuum and the fact that it has not been possible to precisely distinguish cambium from procambium, most workers agree that procambium and cambium are sequential stages of the same meristem (Bond, 1942; Esau, 1943b; Philipson and Ward, 1965; Cumbie, 1963, 1967; Philipson *et al.*, 1971; Fahn *et al.*, 1972; Enright and Cumbie, 1973; Butterfield, 1976; Larson, 1976a).

Swamy and Krishnamurthy (1980) recently questioned the concept of a procambium→cambium continuum. Their principal counterarguments were: 1) interfascicular cambium arises from ground tissue and it does so before the cambium; 2) in roots, the locus of cambial origin bears no relation to the phloic and xylary strands which occur independently; and 3) accessory cambial layers of some species arise in differentiated parenchyma. These authors conclude that procambium is not a prerequisite for cambium and that procambium must first be converted to parenchyma, which in turn gives rise to cambium.

Interfascicular cambium

All cells of the residual meristem do not necessarily contribute to fascicular tissue. Some cells may contribute to interfascicular tissue (Esau, 1943b). When procambial strands are first discernible in the residual meristem, they appear as discrete groups or islands of cells in transverse sections. The cells immediately surrounding and interjacent to the procambial strands have been referred to as interfascicular residual meristem (Sterling, 1946; Larson, 1975). The fate of these cells differs among species and with growing conditions of the plant.

Many cells of the interfascicular residual meristem are incorporated in the developing and enlarging procambial bundles. Many more are incorporated in the subsidiary bundles (Larson, 1975) or accessory bundles (Resch, 1959) that contribute to the leaf trace system. When followed downward in the stem, the remaining cells lying interjacent to the procambial trace bundles become more highly vacuolated and parenchymatous and it is these cells that have been variously interpreted by different investigators.

According to Esau (1943b), the interfascicular meristem is a secondary meristem. Consequently, periclinal divisions that bridge the trace bundles during internodal elongation must be considered part of the procambium. Thompson and Heimsch (1964) concurred with this view and concluded that *Lycopersicon* did not possess an interfascicular cambium because a complete vascular cylinder was established during primary growth. Similarly, Alexandrov and Alexandrova (1929) concluded that *Helianthus* produced no interfascicular cambium. Apparently, the interfascicular region can fill in either by simple enlargement and merger of the procambial bundles as in *Pseudotsuga* (Sterling, 1947) or by periclinal divisions bridging the narrow interfascicular gap early in procambial development. *Populus* provides an example of the latter process (Larson, 1975). During early metacambial development, the procambial bundles, although discrete, are seldom separated by more than two or three cells. This interfascicular gap is bridged by periclinal divisions that advance acropetally and laterally from the fascicular metacambial regions (Fig. 4). Sterling (1946) and Schnettker (1977b) also observed in *Sequoia* and *Helianthus*, respectively, that periclinal divisions united the fascicular bundles in a complete vascular cylinder above the secondary region.

The lateral spread or advance of periclinal divisions from fascicular to interfascicular regions during primary growth is apparently common in many species (Esau, 1943b). Steeves and Sussex (1972) noted that where bundles are widely spaced, periclinal divisions begin adjacent to those in the bundles and advance across the interfascicular regions. They further suggested that the stimulus for cell division originates in the fascicular bundles and spreads to the interfascicular regions (see also Phillips, 1976). Fahn *et al.* (1972) observed a similar pattern of development in *Ricinus,* but they interpreted it differently. Ordered tangential divisions were found to first occur on both sides of the radial rows of procambial derivatives. These divisions extended into the radial row later, shortly before the interfascicular cambium was complete. They therefore questioned the commonly held view that fascicular cambium always precedes interfascicular cambium.

On the basis of experiments with hypocotylary tissue of *Ricinus,* Siebers (1971a) concluded that a direct ontogenetic continuity exists between the primary meristem ring as defined by Kaussmann (see Siebers, 1971b) or the residual meristem as defined by Esau (1965b), and the tissue layer in which interfascicular cambium forms. Consequently, the primary meristem ring represents the predetermined cambium, and both the fascicular and interfascicular parts originate from the same meristematic layer early in primary development (Siebers, 1971b, 1972). On the basis of experiments in which he could detect no induction of interfascicular cambium from fascicular cambium, Siebers (1972) rejected the idea that the interfascicular cambium represents a secondary meristem.

In those species in which periclinal divisions occur late in the interfascicular region, the remainder of the residual meristem eventually undergoes vacuolation and differentiates as interfascicular parenchyma. Derivatives of this tissue form the interfascicular cambium that eventually differentiates during secondary growth in many dicotyledonous and coniferous species (Cutter, 1971). For example, in *Linum* when the fascicular cambium becomes organised within the bundles at the end of primary growth, a cambium also appears in the interfascicular regions (Esau, 1942). In contrast to the inter-

114

ONTOGENETIC SEQUENCE OF CAMBIAL DEVELOPMENT
IN COTTONWOOD

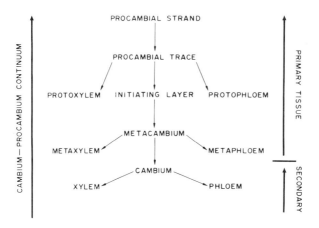

Fig. 10. Proposed ontogenetic sequence of cambial development and origin of derivatives in cottonwood. The cambial-procambial sequence proceeds acropetally and continuously; the arrows have been directed downward to emphasise development. All xylem and phloem derivatives differentiate basipetally and acropetally, respectively. From Larson, 1976a (with permission of the American Journal of Botany).

fascicular divisions that occur early in primary development, those that occur late in development to produce interfascicular cambium are considered to result from de-differentiation of the interfascicular parenchyma (Devadas and Beck, 1971; Phillips, 1976). Once formed, the interfascicular cambium may produce typical vascular tissue as in most woody species or parenchyma as in some vines (Esau, 1943b). Medullary rays also arise in the interfascicular regions (Barghoorn, 1940a).

Conclusion

When viewed ontogenetically, the entire meristematic system of the plant may be considered a continuum. The apical meristems of both shoot and root and the procambium that unites them are derived from early embryonic tissue and perpetuated throughout subsequent plant development. The residual meristem is a transitional tissue between the apical and lateral meristems. As plant development continues, the lateral meristem may be further subdivided into the procambium→metacambium→cambium continuum in those plants that develop a recognisable secondary body. Derivatives of the lateral meristem may also present developmental continua in some plants. Figure 10 represents an attempt by Larson (1976a, 1980a) to subdivide the procambium→metacambium→ continuum in *Populus*. The proposed system may have application to other species.

All the foregoing events are encompassed within the concept of cambium. Although cambium can be distinguished from procambium, the transitional continuum that unites them cannot be clearly identified. Moreover, events that occur during procambial (primary) development largely determine those that occur during cambial (secondary) development. Consequently, procambium must be considered a logical part of the concept of cambium.

Once the cambial state has been attained, the cambium and its derivatives differ from the procambium and its derivatives. The cambium often becomes a complex meristem and its derivatives differentiate to a range of specialised elements. The structure of the cambium and development of its derivatives may be also included in a broad concept of cambium, but they are beyond the scope of this paper. The cambium and its structure have been recently reviewed by Catesson (1974), Tsuda (1975), Phillips (1976), and Timell (1980), and cambial terminology has been discussed in detail by Butterfield (1975) and Schmid (1976). General accounts of the cambium and its development can be found in Romberger (1963), Philipson *et al.* (1971), Kozlowski (1971), and Zimmermann and Brown (1971).

References

Alexandrov, W.G. & O.G. Alexandrova. 1929. Über die Struktur verschiedener Abschnitte ein und desselben Bündels und den Bau von Bündeln verschiedener Internodien des Sonnenblumenstengels. Planta 8: 465-486.

Allen, G.S. 1947. Embryogeny and the development of the apical meristems of Pseudotsuga. II. Late embryogeny. Amer. J. Bot. 34: 73-80.

Allsopp, A. 1964. Shoot morphogenesis. Ann. Rev. Plant Physiol. 15: 225-254.

Amer, M. & P. Neville. 1979. Morphogenèse chez Gleditsia triacanthos L. XIV. Mise en place du cambium caulinaire et influence des organes aériens sur son activité. Rev. gén. Bot. 86: 203-220.

Ans, B. 1979. Formation of a plant primary vascular-like structure in a reaction-diffusion system. Flora 168: 358-378.

Ashworth, R.P. 1963. Investigations into the midvein anatomy and ontogeny of certain species of the genus Ilex. J. Elisha Mitchell Sci. Soc. 73: 126-138.

Bailey, I.W. 1944. The development of vessels in angiosperms and its significance in morphological research. Amer. J. Bot. 31: 421-428.

Balfour, E.E. & W.R. Philipson. 1962. The development of the primary vascular system of certain dicotyledons. Phytomorphology 12: 110-143.

Barghoorn, E.S., Jr. 1940a. Origin and development of the uniseriate ray in the Coniferae. Bull. Torrey Bot. Club: 303-328.

Barghoorn, E.S., Jr. 1940b. The ontogenetic development and phylogenetic specialization of rays in the xylem of dicotyledons I. The primitive ray structure. Amer. J. Bot. 27: 918-928.

Berlyn, G.P. 1967. The structure of germination in Pinus lambertiana Dougl. Yale Univ. School For. Bull. No. 71. New Haven, Conn.

Bisalputra, T. 1961. Anatomical and morphological studies in the Chenopodiaceae. II. Vascularization of the seedling. Aust. J. Bot. 9: 1-9.

Blyth, A. 1958. Origin of primary extraxylary stem fibers in dicotyledons. Univ. Calif. Publ. Bot. 30: 145-232.

Boke, N. 1940. Histogenesis and morphology of the phyllode in certain species of Acacia. Amer. J. Bot. 27: 73-90.

Boke, N. 1944. Histogenesis of the leaf and areole in Opuntia cylindrica. Amer. J. Bot. 31: 299-316.

Boke, N. 1947. Development of the adult shoot and floral initiation in Vinca rosea L. Amer. J. Bot. 34: 433-439.

Boke, N. 1951. Histogenesis of the vegetative shoot in Echinocereus. Amer. J. Bot. 38: 23-38.

Boke, N. 1952. Leaf and areole development in Coryphantha. Amer. J. Bot. 39: 134-145.

Bond, T.E.T. 1942. Studies in the vegetative growth and anatomy of the tea plant (Camelia thea Link.) with special reference to the phloem. I. The flush shoot. Ann. Bot. 6: 607-630.

Bünning, E. 1965. Die Entstehung von Mustern in der Entwicklung von Pflanzen. Handb. Pflanzenphysiol. 15: 383-408.

Butterfield, B.G. 1975. Terminology used for describing the cambium. IAWA Bull. 1975/1: 13-14.

Butterfield, B.G. 1976. The ontogeny of the vascular cambium in Hoheria angustifolia Raoul. New Phytol. 77: 409-420.

Carlquist, S. 1962. A theory of paedomorphosis in dicotyledonous woods. Phytomorphology 12: 30-45.

Carothers, Z. 1959. Observations on the procambium and primary phloem of Pelargonium domesticum. Amer. J. Bot. 46: 397-404.

Caruso, J.L. & E.G. Cutter. 1970. Morphogenetic aspects of a leafless mutant in tomato II. Induction of a vascular cambium. Amer. J. Bot. 57: 420-429.

Catesson, A.M. 1964. Origine, fonctionnement et variations cytologiques saisonnieres du cambium de l'Acer pseudoplatanus L. Ann. sci. nat. Bot. (Ser. 12)5: 229-498.

Catesson, A.M. 1974. Cambial cells, In: A.W. Robards (ed.), Dynamic aspects of plant ultrastructure: 358-390. McGraw-Hill, New York.

Clowes, F.A.L. 1961. Apical Meristems. Blackwell Scientific, Oxford.

Cohen, L.I. 1963. Studies on the ontogeny of the dwarf mistletoes, Arceuthobium I. Embryogeny and histogenesis. Amer. J. Bot. 50: 400-407.

Crafts, A.S. 1943a. Vascular differentiation in the shoot apex of Sequoia sempervirens. Amer. J. Bot. 30: 110-121.

Crafts, A.S. 1943b. Vascular differentiation in the shoot apices of ten coniferous species. Amer. J. Bot. 30: 382-393.

Crooks, D.M. 1933. Histological and regenerative studies on the flax seedling. Bot. Gaz. 95: 209-239.

Cross, G.L. 1942. Structure of the apical meristem and development of the foliage leaves of Cunninghamia lanceolata. Amer. J. Bot. 29: 288-301.

Cumbie, B.G. 1963. The vascular cambium and xylem development in Hibiscus lasiocarpus. Amer. J. Bot. 50: 944-951.

Cumbie, B.G. 1967. Development and structure of the xylem in Canavalia (Leguminosae). Bull. Torrey Bot. Club 94: 162-175.

Cutter, E.G. 1971. Plant Anatomy: Experiment and Interpretation Part 2. Organs. Edward Arnold, London.

Devadas, C. & C.B. Beck. 1971. Development and morphology of stelar components in the stems of some members of the Leguminosae and Rosaceae. Amer. J. Bot. 58: 432-446.

Devadas, C. & C.B. Beck. 1972. Comparative morphology of the primary vascular systems in some species of Rosaceae and Leguminosae. Amer. J. Bot. 59: 557-567.

Dormer, K.J. 1954. The acadian type of vascular system and some of its derivatives I. Introduction, Menispermaceae and Lardizabalaceae, Berberidaceae. New Phytol. 53: 301-311.

Dormer, K.J. 1972. Shoot Organization in Vascular Plants. Chapman & Hall, London.

Eames, A.J. & L.H. McDaniels. 1925. An Introduction to Plant Anatomy. 1st Ed. McGraw-Hill, New York.

Eames, A.J. & L.H. McDaniels. 1947. An Introduction to Plant Anatomy. 2nd Ed. McGraw-Hill, New York.

Enright, A.M. & B.G. Cumbie. 1973. Stem anatomy and internode development in Phaseolus vulgaris. Amer. J. Bot. 60: 915-922.

Esau, K. 1942. Vascular differentiation in the vegetative shoot of Linum. I. The procambium. Amer. J. Bot. 29: 738-747.

Esau, K. 1943a. Vascular differentiation in the vegetative shoot of Linum. II. The first phloem and xylem. Amer. J. Bot. 30: 248-255.

Esau, K. 1943b. Origin and development of primary vascular tissues in seed plants. Bot. Rev. 9: 125-206.

Esau, K. 1945. Vascularization of the vegetative shoot of Helianthus and Sambucus. Amer. J. Bot. 32: 18-29.

Esau, K. 1953. Anatomical differentiation in shoot and root axes. In: W.E. Loomis (ed.), Growth and Differentiation in Plants: 69-100. Iowa State Coll. Press, Ames.

Esau, K. 1954. Primary vascular differentiation in plants. Biol. Rev. 29: 46-86.

Esau, K. 1965a. Vascular Differentiation in Plants. Holt, Rinehart & Winston, New York.

Esau, K. 1965b. Plant Anatomy. John Wiley & Sons, New York.

Esau, K. 1977. Anatomy of Seed Plants. 2nd Ed. John Wiley & Sons, New York.

Esau, K. & I.B. Morrow. 1974. Spatial relation between xylem and phloem in the stem of Hibiscus cannabinus L. (Malvaceae). Bot. J. Linn. Soc. 68: 43-50.

Fahn, A., R. Ben-Sasson & T. Sachs. 1972. The relation between the procambium and the cambium. In: A.K.M. Ghouse (ed.), Research Trends in Plant Anatomy: 161-170. Tata McGraw-Hill, New Delhi.

Frey-Wyssling, A. 1940. Zur Ontogenie des Xylems in Strengeln mit sekundärem Dickenwachstum. Ber. dtsch. bot. Ges. 58: 166-181.

Gifford, E.M., Jr. 1951. Early ontogeny of the foliage leaf in Drimys winteri var. chilensis. Amer. J. Bot. 38: 93-105.

Girolami, G. 1954. Leaf histogenesis in Linum usitatitissimum. Amer. J. Bot. 41: 264-273.

Goffinet, M.C. & P.R. Larson. 1981. Structural changes in Populus deltoides terminal buds and in the vascular transition zone of the stems during dormancy induction. Amer. J. Bot. 68: 118-129.

Griffith, M.M. 1952. The structure and growth of the shoot apex in Araucaria. Amer. J. Bot. 39: 253-263.

118

Gunckel, J.E. & R.H. Wetmore. 1946. Studies of development in long shoots and short shoots of Ginkgo biloba L. I. The origin and pattern of development of the cortex, pith and procambium. Amer. J. Bot. 33: 285-295.

Hara, N. 1975. Structure of the vegetative shoot apex of Cassiope lycopodioides. Bot. Mag. Tokyo 88: 89-101.

Helm, J. 1931. Untersuchungen über die Differenzierung der Sprossscheitelmeristeme von Dikotylen unter besonderer Berücksichtigung des Procambiums. Planta 15: 105-191.

Heslop-Harrison, J. 1967. Differentiation. Ann. Rev. Plant Physiol. 18: 325-348.

Hurka, H. & S. Büchele. 1976. Histologisch-entwicklungsgeschichtliche Untersuchungen am Laubstengel von Capsella bursa-pastoris (Brassicaceae). Flora 165: 369-379.

International Association of Wood Anatomists. 1964. Multilingual Glossary of Terms Used in Wood Anatomy. Konkordia, Winterthur.

Isebrands, J.G. & P.R. Larson. 1973. Some observations on the cambial zone in cottonwood. IAWA Bull. 1973/3: 3-11.

Kaplan, D.R. 1970. Seed development in Downingia. Phytomorphology 19: 253-278.

Kostytschew, S. 1924. Der Bau und das Dickenwachstum der Dicotylenstämme. Beih. Bot. Centralbl. 40: 295-350.

Kozlowski, T. 1971. Growth and Development of Trees. Vol. II. Acad. Press, New York.

Lang, A. 1965. Progressiveness and contagiousness in plant differentiation and development. Handb. Pflanzenphysiol. 15: 409-423.

Lang, A. 1973. Inductive phenomena in plant development. Brookhaven Symp. Biol. 25: 129-144.

Larson, P.R. 1975. Development and organization of the primary vascular system in Populus deltoides according to phyllotaxy. Amer. J. Bot. 62: 1084-1099.

Larson, P.R. 1976a. Procambium vs. cambium and protoxylem vs. metaxylem in Populus deltoides seedlings. Amer. J. Bot. 63: 1332-1348.

Larson, P.R. 1976b. The leaf-cambium relation and some prospects for genetic improvement. In: M.G.R. Cannell & F.T. Last (eds.), Tree Physiology and Yield Improvement: 261-282. Acad. Press, London.

Larson, P.R. 1979. Establishment of the vascular system in seedlings of Populus deltoides Bartr. Amer. J. Bot. 66: 452-462.

Larson, P.R. 1980a. Control of vascularization by developing leaves. In: C.H.A. Little (ed.), Control of Shoot Growth in Trees: 157-172. Proc. IUFRO Workshop, Fredericton, N.B., Canada.

Larson, P.R. 1980b. Interrelations between phyllotaxis, leaf development and the primary-secondary vascular transition in Populus deltoides. Ann. Bot. 46: 757-769.

Larson, P.R. 1982. Primary vascularization and the siting of primordia. In: F.L. Milthorpe & J.E. Dale (eds.), Leaf Growth and Function. Cambridge Univ. Press, Cambridge (In press).

Larson, P.R. & M.C. Goffinet. 1981. Advance of the primary-secondary vascular transition zone during dormancy induction of Populus deltoides. IAWA Bull. n.s. 2: 25-30.

Larson, P.R. & J.G. Isebrands. 1974. Anatomy of the primary-secondary transition zone in stems of Populus deltoides. Wood Sci. Technol. 8: 11-26.

Larson, P.R. & T.D. Pizzolato. 1977. Axillary bud development in Populus deltoides I. Origin and early development. Amer. J. Bot. 64: 835-848.

Little, J.R. & C.E. Jones. 1980. A Dictionary of Botany. Van Nostrand Reinhold Co., New York.

Loy, J.B. 1977. Hormonal regulation of cell division in the primary elongating meristems of shoots. In: T.L. Rost & E.M. Gifford (eds.), Mechanisms and Control of Cell Division: 92-111. Dowden, Hutchinson & Ross, Stroudsburg.

Lyndon, R.F. 1976. The shoot apex. In: M.M. Yeoman (ed.), Cell Division in Higher Plants: 285-314. Acad. Press, New York.

Mahlberg, P.G. 1960. Embryogeny and histogenesis in Nerium oleander I. Organization of primary meristematic tissues. Phytomorphology 10: 118-131.

Majumdar, G.P. 1942. The organization of the shoot in Heracleum in the light of development. Ann. Bot. 6: 49-81.

McGahan, M.W. 1955. Vascular differentiation in the vegetative shoot of Xanthium chinense. Amer. J. Bot. 42: 132-140.

Miller, H.A. & R.H. Wetmore. 1945a. Studies in the developmental anatomy of Phlox drummondii Hook. I. The embryo. Amer. J. Bot. 32: 588-599.

Miller, H.A. & R.H. Wetmore. 1945b. Studies in the developmental anatomy of Phlox drummondii Hook. II. The seedling. Amer. J. Bot. 32: 628-634.

Miller, H.A. & R.H. Wetmore. 1946. Studies in the developmental anatomy of Phlox drummondii Hook. III. The apices of the mature plant. Amer. J. Bot. 33: 1-10.

Millington, W.F. & J.E. Gunckel. 1950. Structure and development of the vegetative shoot tip of Liriodendron tulipifera. Amer. J. Bot. 37: 326-335.

Mogensen, H.L. 1966. A contribution to the anatomical development of the acorn in Quercus L. Iowa State J. Sci. 40: 221-255.

Namboodiri, K.K. & C.B. Beck. 1968a. A comparative study of the primary vascular system of conifers I. Genera with helical phyllotaxis. Amer. J. Bot. 55: 447-457.

Namboodiri, K.K. & C.B. Beck. 1968b. A comparative study of the primary vascular system of conifers II. Genera with opposite and whorled phyllotaxis. Amer. J. Bot. 55: 458-463.

Nast, C.G. 1941. The embryogeny and seedling morphology of Juglans regia L. Lilloa 6: 163-205.

Newman, I.V. 1961. Pattern in the meristem of vascular plants II. A review of shoot apical meristems of gymnosperms, with comments on apical biology and taxonomy, and a statement of some fundamental concepts. Proc. Linn. Soc. N.S. Wales 86: 9-59.

O'Brien, T.P. 1974. Primary vascular tissues. In: A.W. Robards (ed.), Dynamic Aspects of Plant Ultrastructure: 414-440. McGraw-Hill, New York.

O'Neill, T.B. 1961. Primary vascular organization of Lupinus shoot. Bot. Gaz. 123: 1-9.

Parke, R.V. 1963. Initial vascularization of the vegetative shoot of Abies concolor. Amer. J. Bot. 50: 464-469.

Philipson, W.R. & E.E. Balfour. 1963. Vascular patterns in dicotyledons. Bot. Rev. 29: 382-404.

Philipson, W.R. & J.M. Ward. 1965. The ontogeny of the vascular cambium in the stem of seed plants. Biol. Rev. 40: 534-579.

Philipson, W.R., J.M. Ward & B.G. Butterfield. 1971. The Vascular Cambium: Its Development and Activity. Chapman & Hall, London.

Phillips, I.D.J. 1976. The cambium. In: M.M. Yeoman (ed.), Cell Division in Higher Plants: 347-390. Acad. Press, New York.

Pillai, A., S.K. Pillai & O. Jacob. 1975. Embryogeny, histogenesis and apical meristems of Cosmos bipinnatus Cav. Acta Bot. Indica 3: 68-78.

Pillai, S.K. & E.C. Raju. 1975. Some aspects of developmental anatomy of Arachis hypogaea L. Flora 164: 487-496.

Priestley, J.H. & L.I. Scott. 1936. The vascular anatomy of Helianthus annuus L. Proc. Leeds Phil. Soc. 3: 159-173.

Pulawska, Z. 1965. Correlations in the development of the leaves and leaf traces in the shoot of Actinidia arguta Planch. Acta Soc. Bot. Polon. 34: 697-712.

Pyykkö, M. 1974. Developmental anatomy of the seedling of Honkenya peploides. Ann. Bot. Fenn. 11: 253-261.

Ramji, M.V. 1975. Histology of growth with regard to embryos and apical meristems in some angiosperms I. Embryogeny of Stellaria media. Phytomorphology 25: 131-145.

Reeve, R.M. 1942. Structure and growth of the vegetative shoot apex of Garrya elliptica Dougl. Amer. J. Bot. 29: 697-711.

Reeve, R.M. 1948a. Late embryogeny and histogenesis in Pisum. Amer. J. Bot. 35: 591-602.

Reeve, R.M. 1948b. The 'tunica-corpus' concept and development of shoot apices in certain dicotyledons. Amer. J. Bot. 35: 65-75.

Resch, A. 1959. Über Leptombündel und isolierte Siebröhren sowie deren Korrelationen zu den übrigen Leitungsbahnen in der Sprossachse. Planta 52: 467-489; 490-515.

Richards, J.H. & P.R. Larson. 1982. The initiation and development of secondary xylem in axillary branches of Populus deltoides. Ann. Bot. 49: 149-163.

Romberger, J.A. 1963. Meristems, Growth, and Development in Woody Plants. U.S. Dept. Agric. Tech. Bull. No. 1293. Washington, D.C.

Rouffa, A.S. & J.E. Gunckel. 1951. Leaf initiation, origin, and pattern of pith development in the Rosaceae. Amer. J. Bot. 38: 301-306.

Sacher, J.A. 1954. Structure and seasonal activity of the shoot apices of Pinus lambertiana and Pinus ponderosa. Amer. J. Bot. 41: 749-759.

Sachs, R.M. 1965. Stem elongation. Ann. Rev. Plant Physiol. 16: 73-96.

Schmid, R. 1976. The elusive cambium - another terminological contribution. IAWA Bull. 1976/4: 51-59.

Schnettker, M. 1976. Anlage und Differenzierung des Prokambiums im Sprossscheitel von Clematis vitalba (Ranunculaceae). Plant. Syst. Evol. 125: 59-75.

Schnettker, M. 1977a. Die Verteilung der leitenden Strukturen in der Achse von Clematis vitalba (Ranunculaceae). Plant Syst. Evol. 127: 87-102.

Schnettker, M. 1977b. Zum Dickenwachstum bei Helianthus annuus L. (Compositae). Bot. Jahrb. Syst. 98: 250-265.

Schopf, J.M. 1943. The embryology of Larix. Ill. Biol. Monog. 19: 1-97.

Shininger, T.L. 1979. The control of vascular development. Ann. Rev. Plant Physiol. 30: 313-337.

Shushan, S. & M.A. Johnson. 1955. The shoot apex and leaf of Dianthus caryophyllus L. Bull. Torrey Bot. Club 82: 262-283.

Siebers, A.M. 1971a. Initiation of radial polarity in the interfascicular cambium of Ricinus communis L. Acta Bot. Neerl. 20: 211-220.

Siebers, A.M. 1971b. Differentiation of isolated interfascicular tissue of Ricinus communis L. Acta Bot. Neerl. 20: 343-355.

Siebers, A.M. 1972. Vascular bundle differentiation and cambial development in cultured tissue blocks excised from the embryo of Ricinus communis L. Acta Bot. Neerl. 21: 327-342.

Sifton, H.B. 1944. Developmental morphology of vascular plants. New Phytol. 43: 87-129.

Smith, D.L. 1974. A histological and histochemical study of the cotyledons of Phaseolus vulgaris L. during germination. Protoplasma 79: 41-57.

Soh, W.Y. 1972. Early ontogeny of vascular cambium I. Ginkgo biloba. Bot. Mag. Tokyo 85: 111-124.

Soh, W.Y. 1974a. Early ontogeny of vascular cambium II. Aucuba japonica and Weigela coraeensis. Bot. Mag. Tokyo 87: 17-32.

Soh, W.Y. 1974b. Early ontogeny of vascular cambium III. Robinia pseudo-acacia and Syringa oblata. Bot. Mag. Tokyo 87: 99-112.

Spurr, A.R. 1949. Histogenesis and organization of the embryo in Pinus strobus L. Amer. J. Bot. 36: 629-641.

Stebbins, G.L. 1974. Evolution of morphogenetic patterns. Brookhaven Symp. Biol. 25: 227-243.

Steeves, T.A. & I.M. Sussex. 1972. Patterns in Plant Development. Prentice-Hall, Englewood Cliffs, New Jersey.

Sterling, C. 1945. Growth and vascular development in the shoot apex of Sequoia sempervirens (Lamb.) Endl. II. Vascular development in relation to phyllotaxis. Amer. J. Bot. 32: 380-386.

Sterling, C. 1946. Growth and vascular development in the shoot apex of Sequoia sempervirens (Lamb.) Endl. III. Cytological aspects of vascularization. Amer. J. Bot. 33: 35-45.

Sterling, C. 1947. Organization of the shoot of Pseudotsuga taxifolia (Lamb.) Britt. II. Vascularization. Amer. J. Bot. 34: 272-280.

Sterling, C. 1949. The primary body of the shoot of Dianthera americana. Amer. J. Bot. 36: 184-193.

Swamy, B.G.L. & K.V. Krishnamurthy. 1980. On the origin of vascular cambium in dicotyledonous stems. Proc. Indian Acad. Sci. (Plant Sci.) 89: 1-6.

Thoday, D. 1922. On the organization of growth and differentiation in the stem of the sunflower. Ann. Bot. 36: 489-510.

Thompson, N.P. & C. Heimsch. 1964. Stem anatomy and aspects of development in tomato. Amer. J. Bot. 51: 7-19.

Tilton, B.R. & B.F. Palser. 1976. Primary xylem maturation in conifer seedlings. Bot. Gaz. 137: 165-178.

Timell, T.E. 1980. Organization and ultrastructure of the dormant cambial zone in compression wood of Picea abies. Wood Sci. Technol. 14: 161-179.

Tsuda, M. 1975. The ultrastructure of the vascular cambium and its derivatives in coniferous species I. Cambial cells. Bull. Tokyo Univ. For. 67: 158-226.

Tucker, S.O. 1962. Ontogeny and phyllotaxis of the terminal vegetative shoots of Michelia fuscata. Amer. J. Bot. 49: 722-737.

Vaughan, J.G. 1955. The morphology and growth of the vegetative and reproductive apices of Arabidopsis thaliana (L.) Heynh., Capsella bursa-pastoris (L.) Medic. and Anagallis arvensis L. Linn. Soc. London J. Bot. 55: 279-301.

Wardlaw, C.W. 1965. The organization of the shoot apex. Handb. Pflanzenphysiol. 15: 966-1076.

Wareing, P.F. 1978. Determination in plant development. Bot. Mag. Tokyo (Special issue) 1: 3-17.

Warren-Wilson, J. & P.M. Warren-Wilson. 1981. The position of cambia regenerating in grafts between stems and abnormally-oriented petioles. Ann. Bot. 47: 473-484.

Wetmore, R.H. 1943. Leaf-stem relationships in the vascular plants. Torreya 43: 16-28.

Wetmore, R.H., A.E. DeMaggio & J.P. Rier. 1964. Contemporary outlook on the differentiation of vascular tissues. Phytomorphology 14: 203-217.

Wetmore, R.H. & T.A. Steeves. 1971. Morphological introduction to growth and development. In: F.C. Steward (ed.): Vol. 6A: 3-168, Plant Physiology: A Treatise. Acad. Press, New York.

White, D.J.B. 1955. The architecture of the stem apex and the origin and development of the axillary bud in seedlings of Acer pseudoplatanus L. Ann. Bot. 19: 437-449.

Winter, C.W. 1932. Vascular system of young plants of Medicago sativa. Bot. Gaz. 94: 152-167.

Yeung, E.C. & M.E. Clutter. 1978. Embryogeny of Phaseolus coccineus: Growth and micro-anatomy. Protoplasma 94: 19-40.

Young, B.S. 1954. The effects of leaf primordia on differentiation in the stem. New Phytol. 53: 445-460.

Zimmermann, M.H. & C.L. Brown. 1971. Trees Structure and Function. Springer, Berlin.

Morphogenetic factors in wood formation and differentiation

GRAEME P. BERLYN

Yale University, School of Forestry and Environmental Studies, New Haven, Connecticut 06511, U.S.A.

Summary: The cambium is discussed and analysed in relation to hypotheses that consider the cambium to be a morphogenetic organiser and those that consider the cambium to be merely a cell maker that is entirely controlled by external influences. The question of whether the cambium is a single tier of initials or a multiseriate meristem is investigated in regard to its physiological and morphogenetic basis. The cambium undergoes three different types of cell divisions, *viz.* multiplicative, additive, and transformative. Multiplicative divisions result in new initials that enable the cambium to keep pace with the increase in girth of the stem while additive divisions are tangential-longitudinal divisions that give rise to xylem and phloem derivatives. Transformative divisions are those whereby the fusiform initial divides up to become a ray cell initial (or initials). These different types of division arise from altered planes of the mitotic spindle. The factors controlling these divisions are reviewed in relation to their space-time elements and their correlation of the overall growth pattern of the tree. The cambial stimulus and resulting morphogenetic domains are examined along with biophysical factors in relation to their effect on cambial activity, orientation and qualities of the cambial derivatives.

Preamble

According to the late E.W. Sinnott (1960) morphogenesis is concerned with the causal aspects of organic form and thus includes aspects of genetics, morphology, physiology and embryology. This paper will consider the factors involved in determining the number, types and qualities of the xylem cells formed by the cambium. In order to approach this problem, I will first summarise present concepts of the cambium especially as they have developed over the last fifty years. The ontogenetic origin of the cambium will not be considered as this will be dealt with by Philip Larson in this volume. Extensive coverage of reaction wood and xylem cell walls will also be omitted as these topics have been the subject of numerous reviews (see Berlyn, 1979; Timell, 1973; Wardrop 1964; Westing, 1965, 1968).

An important morphogenetic question is whether the cambium is a morphogenetic organiser or whether it merely makes cells whose fates are determined by extrinsic factors. Of course, the cambium is not a *tabula rasa* that produces cells whose destiny is totally controlled by the environment. The fact that woods of different species contain diagnostic features that enable them to be identified is some evidence on this point. Furthermore, Worrall (1980) describes some reciprocal grafts with hard pines and soft pines that showed that a soft pine stock continues to produce soft pine wood even when attached to a hard pine scion and *vice versa*. Also, in tissue culture, pine stem segments continue to produce their genetic type of wood despite being disconnected from a crown and artificially supplied with nutrients and hormones (see Berlyn and Beck, 1980). Obviously there is a specific range of response for each genotype and the phenotype produced is itself a manifestation of morphogenesis. On the other hand, the cambium cannot be totally autonomous because it depends on the leaves, to which it is connected by the cambium-metacambium-procambium-continuum, for at least photosynthate, growth regulators, and time signals of various kinds.In addition, the cambium is dependent on the roots for growth regulators, water and minerals and possibly other substances. If these dependencies mean the cambium cannot be totally autonomous, they also mean that it is coordinated throughout the entire tree and this coordination is also a manifestation of morphogenesis. There is of course the distinct possibility that the cambium and the developing xylem and phloem have the capacity to synthesise their own growth regulators but at the very least the cambium is dependent on the leaves for carbon and on the roots for minerals and water.

The cambium develops from tissues formed by the apical meristems but it is heterogeneous in origin in that it can form in procambium, parenchyma, callus, or residual meristem. Once formed the cambium is fundamentally different in organisation from the apical meristems. It consists of a single tier of cells, it gives rise to specific tissues and not whole organs, and from an evolutionary point of view it is an optional meristem. While every vascular plant requires at least a shoot meristem, the cambium is not universal. In the phyletic line some groups have both gained and lost a cambium (see Barghoorn, 1964). There is also developmental variation, *e.g.* in some annual and even perennial dicots a cambium forms but it functions for only a limited time span. At present we do not know what causes either phylogenetic or morphogenetic terminations.

Although there are opposing viewpoints (Catesson, 1964, 1980a,b; Murmanis, 1971), the preponderance of evidence supports the concept that the cambium is a single tier of cells (see Timell, 1980a,b). The basis for this is:

(1) The radial continuity of cells across the cambium; the generally simultaneous origin of new radial files and simultaneous cessation of old ones in both the xylem and phloem.

(2) The observation that increase in girth of the xylem is accompanied by increase in girth of the phloem (a consequence of the simultaneous origin of new files).

(3) The fact that slight variations in cell shape in a radial file in the xylem can be traced back to the phloem where comparable modifications have occurred.

A spirited discussion of this point is given by I.W. Bailey and A.B. Wardrop in Zimmermann (1964, pp. 34-35). They supported the rigorous use of the word cambium to refer only to the fusiform and ray initials, but they somewhat grudgingly agreed that the zone of undifferentiated cells between the xylem and phloem could be called the cambial zone.

The vascular cambium consists of initials only − cells that give rise to other cells. In general, when an initial divides one of the daughter cells retains the initial status while the other becomes directly or indirectly a xylem or phloem cell. In the usual case, the non-initial cell produced is termed a xylem mother cell (XMC) if it is on the xylem side or a phloem mother cell (PMC) if it is on the phloem side; however, many variations occur especially in species with included phloem or multiple cambia (see Bormann and Berlyn, 1981 for some of the variations in tropical species). The XMC (and the PMC) may divide one or more times before they are fixed on a path of terminal differentiation. Some modern workers still adhere to Sanio's view (1873) that these divisions occur in pairs under close morphogenetic control (see Timell, 1973). It is difficult to subscribe to a determinism as rigid as that described by Timell (1980a) which requires that the dormant cambial zone consists almost exclusively of radial files each of which contains a group of cells called Sanio's four. This group consists of the initial, a mother cell, and two daughter cells. In *Vitis* Esau (1948; 1977, p.49) states that during dormancy the cambial zone may be reduced to a single cell in width and that these cells comprise the true cambium. Thus at least in this case there is no 'Sanio's four.'

A major question about the cambium is how a meristem that consists of only two types of initials (the ray initials which give rise to the radial system of the tree and the fusiform initials which give rise to the axial system) can produce wood with many different types of cells (vessels, tracheids, fibres and parenchyma cells of various sorts). However, there is *some* variation among the initials of each type. For example, the ray initials are often described as isodiametric cells, but Wodzicki and Brown (1973) showed that in some species of *Pinus* the ray initials often vary in their dimensions and in general consist of two types:

(1) Marginal ray initials whose radial diameters are equal to or less than their axial lengths, and

(2) Interior ray initials whose radial diameters are greater than their axial lengths.

The proportion of ray initials in the cambium may vary from 0 to 40% but it is most commonly in the range of 10-20% (see Paliwal and Srivastava, 1969; Ghouse and Yunus, 1974, 1976.) Also, in the Pinaceae the cambium may be interrupted by radial resin ducts (Wodzicki and Brown, 1973).

Despite the fact that the cambium is a single tier, it is sometimes difficult to identify the true initial in each radial file of cells in the cambial zone (Timell, 1980a). Nevertheless, the initial cells, whether fusiform or ray initials, are the only cells that have the morphogenetic capacity to produce cells in both a centripetal and centrifugal direction (additive divisions); the fusiform initials are with rare exceptions the only cells that divide anticlinally (multiplication divisions). However, when additive cell divisions begin in the spring (or at more frequent periods in some tropical species) the initials may not be lined up because of asynchrony in the cell cycles of the initials (Newman, 1956). This is particularly common in vessel-bearing species where the fusiform initial gives rise to a vessel element (or elements, in the case of nested vessel species) and then enters a quiescent phase until a morphogenetic time signal unblocks the cell cycle and permits further additive divisions. Newman (1956) reported that in *Pinus radiata* the fusiform initial has a xylem-forming and phloem-forming phase. Every time the initial divides, a complete wall layer is formed around the protoplast (Wardrop, 1952). To locate the fusiform initial one first finds a ray initial (easy to locate because of its larger radial diameter) and then in this general area identifies the fusiform initial as the cell in the cambial zone that has one thick tangential wall. If the initial is in a xylem-forming phase, the thick wall will be on the phloem side and conversely if the initial is in a phloem-forming phase the thick wall will be on the xylem side. This observation has been supported by Mahmood (1968, 1971) and has been extended to *Quercus rubra* (Murmanis, 1977) and *Populus deltoides* (Isebrands and Larson, 1973). However, Catesson and Roland (1981) have recently disputed the emboxing concept of cell wall formation by the cambium. This calls into question the basis for the Newman type analysis. Timell (1980a) lists five techniques for identifying initials and there are probably several more that could be used.

If one carefully examines the cell walls of both active and dormant cambial initials with a good polarising microscope and sufficient light intensity the cell walls, both radial and tangential, will exhibit birefringence (Bailey and Kerr, 1934). The anisotropy of the cambial cell walls is due to the presence of crystalline cellulose. It is not merely a consequence of rodlet or platelet birefringence, since the walls are birefringent even when saturated with a liquid having the same index of refraction as the cell walls (Bailey and Kerr, 1934). Recently several authors have reported that the radial walls of active (but not dormant) fusiform initials contain considerably less cellulose than the tangential walls (Roland, 1978; Catesson, 1980b). If this is indeed the case, then an interference microscope or a polarising microscope equipped with a light-sensing device (cytophotometer) should readily locate the active initials in cross section. These procedures combined with DMSO, pectinase, and hemicellulase extraction should provide a relatively easy and reliable method of localisation of fusiform initials (see Berlyn and Miksche, 1976, p. 159-168, 278-279 for techniques). However, it should be recognised that the apparent thinness of cambial cell walls in both light and electron micrographs may be a consequence of preparative procedures; fresh, undehydrated walls are much thicker (Bailey and Kerr, 1934). The asynchrony of the initials means that the initials form a wavy line around the circumference of the tree, in effect a horizontal wave. However, since most tree trunks are essentially round and smooth the wave is coordinated so that it moves around the trunk at a relatively uniform rate. Secondly, the initials located in proximity to each other tend to function as a group (intercorrelated population) in that their derivatives resemble each other in morphology and chemistry (Berlyn, 1961, p. 374-379). Ford and Robards (1976) noted that there is autocorrelation within a single radial file but did not extend their analysis to neighbouring files. This intercorrelation suggests that there is a morphogenetic clumping in cambial function that is coordinated horizontally to effect relatively uniform diameter growth over time. The coordination also has to extend vertically but the extent of this has not been adequately studied. A cytochemical adaptation of the optical shuttle technique developed by Zimmermann and Tomlinson (1966, 1967) might be useful in this regard.

The outlines of cambial cytology were known to Hartig (1901); however, it was the pioneering researches of I.W. Bailey that brought attention and full elucidation of the phenomenon of additive and multiplicative cell division in the cambium (Bailey, 1919, 1920a, 1920b, 1920c, 1923, 1930; Bailey & Kerr, 1934). The fusiform initials of gymnosperms and of many dicots are elongated cells that form an unstratified sheath around the stems and branches. In more specialised dicotyledons the fusiform initials are generally shorter and picket-shaped and may even form a stratified (storied) sheath around the bole and branches. The cell walls are relatively thin, unlignified, and unique in being so perforated with primary pit fields that the walls take on a beaded appearance in tangential sections. Another unique feature of these cells which may be relevant to their morphogenetic function is their vacuome. In most species the dormant fusiform initials contain vacuoles that are numerous and relatively small (Bailey, 1930). However, in the active state the vacuoles aggregate into one or two large vacuoles which are traversed by anastomosing cytoplasmic strands. The parietal cytoplasm with its numerous mitochondria and lipid bodies undergoes rapid protoplasmic streaming (cyclosis). With respect to the length of the cells, the velocity of movement, the quantities of materials moved, and the size of the large stationary nucleus, this cyclosis is unique, however, its functional significance is not known. Bailey (1930) reported that the tonoplast membrane appeared to move with the protoplasmic stream, and this might be taken as supporting evidence for the fluid mosaic model of membrane structure. The cyclosis is not interrupted during cell division but does cease at low temperatures ($< 0°C$). In some species, *e.g. Fraxinus americana,* the large vacuole is present in both the dormant and active state whether cyclosis is occurring or not (Bailey, 1930). The ray initials are also conspicuously vacuolated but the pattern varies greatly from cell to cell even in the same individual. One of Bailey's important contributions was the study of living initials which he endeavoured to observe within 60 seconds after sectioning. More work along these lines is needed.

According to Stewart (1957,1966) and others, the cambial protoplast is a complex mixture of organic compounds including carbohydrates, lipids, amino acids, proteins, organic acids, various nitrogenous substances, inorganic compounds, and growth regulators. Carbohydrates are the chief component and sucrose is the most abundant carbohydrate followed by glucose and fructose. The proportional representation of these compounds varies with time of year, physiological condition, and species.

In recent years, a number of ultrastructural studies have extended our knowledge about the structure of the cambial cytoplasm (Srivastava, 1966; Srivastava and O'Brien, 1966; Robards and Kidwai, 1969; Mia, 1970; Itoh, 1971; Barnett, 1973, 1975; Timell, 1973, 1979, 1980a; Tsuda, 1975; Farooqui and Robards, 1979). Active cambial initials may differ from initials of apical meristems in having a large and complex vacuome, but the dense cytoplasm is very similar to that of apical initials. The essentially parietal cytoplasm contains abundant mitochondria, plastids, peroxisomes, ribosomes, lipid bodies, golgi bodies, microtubules, endoplasmic reticulum (rough and smooth), amyloplasts, and numerous lomosomes (Murmanis, 1970; Timell, 1973; Catesson, 1980a). The plastids appear to be chloroplasts, but the thylakoid membranes are not aggregated into grana stacks. They are very similar to chloroplasts found in roots and hypocotyls of germinating seedlings under low light conditions (Berlyn, 1970). These plastids in active cambial initials and derivatives often contain starch grains. The origin and allocation of this starch should be of considerable interest. Protein bodies and ergastic substances may also be present in the initials, particularly in the dormant state. The cytoskeleton of the cambium has yet to be investigated (see Berlyn, 1979), but in view of the intense synthetic activity and cytoplasmic streaming of these cells, it is probably well developed (see Parthasarathy and Pesacreta, 1980).

Bailey (1919, 1920a, 1920d) made extremely detailed and meticulous studies of the nuclear morphology of fusiform initials during mitosis and cytokinesis. However, the cell cycle condition of the cambial nucleus has not been adequately investigated. Catesson (1980a) postulates that the cambial nuclei are in the Gl stage during dormancy because this has been reported to be the case for shoot apical meristems in a number of woody species (Owens and Molder, 1973; Cottignies, 1979). However, the apical meristems of dormant pine embryos were found to be at the 3C level of cellular DNA which would suggest they were in S-phase (Dhillon *et al.*, 1978; Berlyn *et al.*, 1979; Patel and Berlyn, 1981a). Since cambial initials differ in structure and physiology from other initials the question of cell cycle state in dormant cambial nuclei must await specific studies. Special features of the nuclei of the fusiform initials are their ellipsoidal shape and their several large nucleoli. Phloem mother cells retain nuclear dimensions that are similar to those of the cambium but developing xylem tracheids have nuclei that are greatly elongated in comparison with those of the cambium. Elongated nuclei also occur in the protoxylem (Berlyn, 1972, p. 284-285) and these huge nuclei develop a complex structure with a diffuse vacuome of their own.

The nucleoli of initials are not only large and numerous, but they have a complex structure which may contain vacuoles filled with both amorphous and birefringent material. Bailey thought the large nucleoli were aggregates of smaller ones but this has not been substantiated. I have observed nucleolar-like material in the egg cytoplasm of *Pinus lambertiana* adjacent to the nucleus. Superficially it appeared that small nucleoli had been extruded into the cytoplasm, but a developmental study suggested that the material was first deposited into the nucleolar vacuoles. After the nucleolus assumed a position along the nuclear membrane, the vacuolar material appeared to move into the cytoplasm. In general, ray initials have a smaller, more spherical nucleus than fusiform initials and the nuclei have fewer nucleoli.

Catesson (1980a) investigated the seasonal cycle in ray initials (and possibly their immediate derivatives since she does not use a strict definition of cambium). In these cells she found that nucleolar volume was at a maximum in April and a minimum in October. The nucleoli had granular and fibrillar phases and also contained the vacuoles which she termed lacunae. In late summer the granular phase gradually decreased until it became reduced to a narrow peripheral zone; at this time the main body of the nucleolus was fibrillar with a few small lacunae (Catesson, 1980a).

We know so little about the cambial genome that at this point it does not seem useful to speculate about its genetic structure or on possible molecular mechanisms of genetic control. This is not meant to imply that this is not a fruitful area for future research, we simply do not have any data on these topics at present. The cytology of mitosis was worked out in detail by Bailey (1919) and no substantial additions have been made since then. The kinetics of mitosis and cytokinesis have been investigated by Wilson (1963, 1964, 1970), Wilson and Howard (1968) and Kennedy and Farrar (1965). M-phase of the cambial cell cycle takes from four to six hours which is about the same time that most plant cells take for both mitosis and cytokinesis. However, in fusiform initials it takes an additional 18-20 hours to complete cytokinesis because the cell plate and new cell wall are greatly extended in length and mass because of the tremendous volume of the fusiform initials (up to $5 \times 10^6 \ \mu m^3$). Thus the present evidence suggests that it takes *ca.* 24 hours for the fusiform initial to produce a new derivative. The literature suggests that when growth rate increases, the frequency of division of each fusiform initial increases but the velocity of cell division is invariant. This concept is somewhat difficult to accept since there are very few invariants in biology. However, if true, this invariance could have important consequences in forest management for optimal wood production.

Multiplicative (anticlinal) divisions

Bailey's classic paper of 1923 clarified the controversy created by earlier workers such as Nägeli, Hartig, and Sanio about the increase in girth of the cambium and provided the basis for the modern concept. He demonstrated that the increase in girth of the cambium was primarily due to an increase in the number of cambial initials and not to the increased tangential diameter of the initials. This was true despite the fact that the tangential diameter of white pine fusiform initials increased 2.6 fold in 60 years from seedling to early maturity. Simultaneously the circumference of the cambium was increasing 100 fold. The combined increase in circumference of the cambium produced by the increased tangential diameters of the cambial initials was 10,032 μm while the actual increase in circumference in 60 years was 1,244,074 μm. In the 60 year interval, the number of fusiform initials increased from 724 to 23,100 and the number of initials increased from 70 to 8,796. (The ray initials only increased 1.2 fold in tangential diameter). A calculation from Bailey's data shows that the ratio of fusiform initials to ray initials went from 10.3 at age one to 2.6 at age sixty. If one puts this on a circumference basis by multiplying the number of initials by their average tangential diameter, the ratio change is more modest, *i.e.* from 11.8 to 6.5. The fusiform initials also increased in length from 870 μm to 4000 μm and since in this species the fusiform initials are unstoried, a correction for the additional cells in a given 10 μm cross section is required. However, the corrected ratio is 6.6 so this did not cause much difference in the 1.8 fold reduction in the ratio. Therefore, by several measures, the older fusiform initials have closer proximity to a ray than do the younger fusiform initials. This of course could be related to increased requirements for cell wall building blocks (carbon) and other metabolites that may be supplied by rays. It should be noted that in many cases, the tangential diameters of the fusiform initials observed over the life of a tree increased more than the 2.6 fold observed by Bailey; however, the significance of Bailey's data remain unchanged.

In highly specialised angiosperms having a storied cambium (*e.g. Robinia pseudoacacia*) the fusiform initials divide in essentially a radiolongitudinal plane although the division plane is usually a little acentric. In other angiosperms and conifers (where the fusiform initials are unstoried) the division, although showing wide variation, is essentially transverse − termed pseudotransverse or pseudoanticlinal. In the latter case, the two daughter initials will then grow past each other by sliding growth and apical intrusive growth. Statistically, these multiplicative divisions add to the girth of the cambium but individually the fate of the new initials is variable. One or both of a given pair may be extruded from the cambium (Bannan, 1950, 1951, 1953, 1956, 1957a, 1957b, 1967, 1968a, 1968b, 1970; Bannan and Bindra, 1970; Whalley, 1950; Hejnowicz, 1961). Elongating initials with contacts to numerous ray initials are the ones most likely to survive as initials. In general, when a fusiform initial is formed it requires about 60 years to achieve its maximum (genetic) length. Since the fusiform initial has a finite probability of undergoing an anticlinal division, the population of initials in unstoried species has a dispersion of initial length which is reflected in the high variance associated with fibre length. Anticlinal divisions are usually restricted to the fusiform initials but XMC's also have the capacity to divide anticlinally and occasionally may do so.

Fibre length is not a simple function of initial length even in conifers. Bannan and Bindra (1970) have shown that cell length, especially during the middle to late growth of the stem, is more closely and inversely related to rate of radial growth than to frequency of anticlinal division of the fusiform initials. In the early stages of seedling and tree growth, cell length and initial length are both short and the rate of anticlinal division of the initials is high; as additional growing sheaths are added to the stem, cell length increases and the rate of anticlinal division declines. Bannan (1970) concluded after studying numerous species of conifers that anticlinal divisions were frequent when radial growth was very low ($\leqq 0.5$ mm yr^{-1}) and decreased rapidly as radial growth increased. The anticlinal divisions stabilised when radial growth was about 2 mm yr^{-1} and thereafter remained constant and infrequent (2 mm^{-1} of xylem) as radial growth increased over a wide range ($\geqq 4.5$ mm yr^{-1}). On the other hand, cell length increased as radial growth went from *ca.* 0.5 mm yr^{-1} to 1 mm yr^{-1} and thereafter cell length decreased continuously as radial growth went from 1 mm yr^{-1} to 4.5 mm yr^{-1}. These data account for a number of conflicting reports in the literature. When growth conditions are suboptimal resulting in radial growth of 0 to 1 mm yr^{-1}, the tree cannot fully express its morphogenetic pattern or characteristics. Thus, for trees sampled under these conditions a set of data could indicate that cell length increased with radial growth rate. This of course would be a proverbial half-truth, *i.e.* true under certain limiting conditions.

From a morphogenetic point of view a major question is what induces these multiplicative divisions of the cambium? Some of the factors involved may be: (1) mechanical stress induced by the increasing circumference of the tree; (2) biofeedback from the cambial derivatives; or (3) a combination of the above. Proximity to rays is also a factor since Bannan (1962) found that the frequency of multiplicative divisions was 16% greater in fusiform initials adjoining rays.

Several studies have shown that the cambium requires pressure in order to function in additive divisions (Brown and Sax, 1962; Brown, 1964) and it has also been observed (Berlyn, 1963) that heat cracks adjacent to ray parenchyma permit proliferation of the ray parenchyma into the crevice of the crack. The cells that filled the crevice were horizontally oriented tracheids complete with circular bordered pits. Thus mechanical forces can induce morphogenesis and certainly tensile and perhaps some shear forces are created in the cambium as diameter growth occurs. However, the fact that the frequency of multiplicative divisions actually declines as growth rate (additive divisions) exceeds 1 mm yr^{-1} and then assumes a constant rate suggests that mechanical forces are not acting alone. This additional factor may be some sort of biofeedback from the developing derivatives or ray initials. Possibly as the derivatives reach a certain number they produce sufficient quantities of an inhibitor that limits multiplicative but not additive divisions. This inhibitor could be involved with the programmed autolysis of the cytoplasmic contents of the tracheary tissue. Another possible mechanism could involve repression of genes whose products could also be repressors which combine with genes that promote specific cell divisions. The regulation must ultimately act on the plane of cell division and thus involves the microtubular and other cytoskeletal elements (microfilaments, intermediate fibrils) since they are involved with the formation and orientation of the spindle.

Once a given fusiform initial undergoes a multiplicative division, the daughter initials elongate rapidly despite the fact that one or occasionally both daughter initials may be lost from the cambium. Since in a single radial file there can be as many as four multiplicative divisions in a single year (Bannan, 1950), there is considerable heterogeneity in length of initials and their derivatives in unstoried species. The rapid elongation is associated with both sliding (gliding) growth and apical intrusive growth. The initials have the capacity to twist around which is quite rare in plants. The adjustments that are made in the cambial cell wall are remarkable. Pit fields are continuously formed and dissolved as wall contacts fluctuate. In some cases, the elongation of the initials appears to be polar (basipetal from a branch tip) (Bannan, 1956) and in other cases (Evert, 1961) the elongation occurs both acropetally and basipetally. Basically, the initials elongate into the available space; if an initial is lost from the cambium a space is available and an inclining initial will elongate into it rapidly. As discussed previously elongation of the initials provides for some of the increase in girth of the cambium and this normally occurs at a relatively low rate unless there are sudden inputs of space. There are far more multiplicative divisions than are needed for circumferential expansion. This overproduction then is followed (in a typical Darwinian manner) by developmental selection. Cell length at the termination of the multiplicative division and proximity of rays appear to be the key elements in this survival of the fittest. Of the two factors ray contact appears to be the most important. Rarely is a large healthy fusiform initial abruptly ejected from the cambium; instead the initial gradually decreases in size (declines, fails) until it reaches some minimal size whereupon it is lost from the cambium leaving only the record of its progressive demise etched in the xylem.

Additive divisions

Additive divisions are the tangential-longitudinal divisions that add cells to the wood body. They either directly or indirectly determine the amount, proportion and type of wood elements, namely vessels, fibres, tracheids, longitudinal parenchyma, resin ducts, and ray tissue. These characteristics are diagnostic for a species but again the expression of the traits, within the limits of phenotypic plasticity, is a morphogenetic process. Growth regulators produced in the leaves (and roots) can only operate on the cambium within these restrictions. Furthermore, a population of cambial initials at a given space-time point (say north facing breast height at 60 years) is not operating in a vacuum. Diameter growth is correlated with height growth and crown dimensions (Berlyn, 1962) and in some species there is an overall pattern of height growth during the life span of the tree. For example, in mixed oak-maple-birch stands in the central New England region of the United States, red oak (*Quercus rubra*) generally grows more slowly than its competitors during the first two decades of its life. However, later in the development of the stand the red oak emerges above the previously dominant canopy (Oliver, 1978, 1980). As this occurs, wood formation is greatly accelerated (Oliver, 1978, p. 41). Such features of the life cycle of the tree must be considered if we are to understand all of the factors affecting wood formation. Thus the additive cell divisions leading to wood production cannot be considered apart from the xylem cell differentiation (XCD) system of which they are an integral part. Berlyn (1970, 1979) identifies six overlapping phases of XCD, *viz.* (1) karyokinetic phase – includes the presynthetic stage (G_1), and the DNA synthesis stage (S), the post synthetic stage (G_1), and mitosis (M); (2) the stage of phragmoplast movement (disk, halo, and frame); (3) cell enlargement and continued primary cell wall formation; (4) cell wall thickening; (5) lignification; and (6) cytoplasmic senescence, autolysis, and elution. Not all xylem cells undergo all six phases but all are created by a division phase and all eventually die. The karyokinetic phase is not restricted to the cambial initial as each xylem derivative has the capacity to divide one or more times before the karyokinetic capacity is suppressed. Consequently 8 to 12 cells of a given radial file of differentiating xylem cells may be in stages 1 and 2 during periods of rapid growth. Fosket (1968, 1970, 1972; Fosket and Torrey, 1969) has postulated that there is a close relationship between a cycle of DNA synthesis and xylogenesis but the nature of this relationship has not been elucidated (Roberts, 1969, 1976). In general the literature suggests that XCD is *usually* preceded by cell division or at least DNA synthesis. Roberts (1976, p. 53) proposes that XCD requires specialised proteins and that these synthetic demands, depending on the species, are met by (1) normal cell cycle; (2) DNA elevation due to endoreduplication or polyteny; (3) gene amplification; or (4) some other special type of DNA production. At any rate inhibitors of DNA synthesis like 5-fluorodeoxyuridine and mitomycin C have also been found to inhibit XCD.

The enzymatically catalysed phases of differentiation which result in the assembly of xylem cells are powered directly by ATP energy, but the ultimate energy source is of course the negentropy of the sun that flows through the leaves, and in variously transformed manifestations, into the sites of xylogenesis. Some of this energy is in the form of information because the cybernetic system that controls the wood formation process requires information to be operational. Some fundamental properties of the process are: (1) it is open – energy and material are exchanged with the environment; (2) it exhibits phenotypic plasticity – within limits it can respond to a variety of environmental stressors; and (3) it is homeostatic – it tends to exist in a dynamic steady state with feedback controls.

One of the classic papers on additive divisions is that of Kennedy and Farrar (1965) which deals with *Pinus banksiana* and *Larix laricina*. Sequentially induced arcs of compression wood (CW) were used to determine the time required for cell division by the cambium and for the stages of cell enlargement, wall thickening, and lignification during xylem differentiation. The two sequential CW arcs served to delimit cells formed during the time interval between arcs. Under the growth conditions prevailing in these experiments the cambial zone consisted of 20 cells. The cambium formed 1 cell per day and 1 cell per day 'completed' differentiation (lignification was the criterion used by these authors). Thus 20 days were necessary for 'all differentiation phases to be completed'. The first 10 days comprised the phase of cell division and in each radial file there were 10 cells in this phase – 1 initial and 9 xylem mother cells, each at least potentially capable of cell division. A limited amount of radial and longitudinal enlargement also occurs in this phase. The phase of enlargement required 2 to 3 days and the phases of cell wall thickening and lignification each required an additional 4 days. While overlap probably occurs between these phases (Wardrop, 1957) the time figures are quite similar to those found by other workers (Dadswell, 1963; Wilson, 1963, 1964).

The phase of lignification does not signal the completion of tracheid differentiation. At this point the tracheid reaches full size and therefore achieves maximum function in mechanical support. However, full maturation includes death and elution of the protoplast; these cells are programmed to be dead and empty at full functional maturity, serving in both water transport and mechanical support. In tracheids this phase of cytoplasmic autolysis requires an additional 2-5 days (Cronshaw, 1965; Cronshaw and Bouck, 1965; Skene, 1972; Wodzicki and Brown, 1973). Skene (1972) reported that this time requirement (4 days in his study) did not change with tree vigor or time of year except that very late in the year time for autolysis increased as did the time for all the other phases of differentiation. Interestingly, Wodzicki and Brown (1973) report that the phase of autolysis begins abruptly with little prior cytoplasmic alteration. The cytoplasm of the tracheids mirrors that of the cambium. A peripheral portion and a reticulum of cytoplasmic strands penetrates the central vacuole. These authors note that such a system provides for rapid exchange of nutrients and metabolites with the vacuolar sap. This system is thought to be essential for the phases of wall thickening (and probably lignification). The origin of the materials in the lomosomes thought to be associated with wall formation (Berlyn, 1970) may well be in this system. However, Wodzicki and Brown (1973) were not able to demonstrate lysosomes in the tracheids although they did observe them in nearby ray parenchyma. At the conclusion of lignification the cell abruptly underwent protoplasmic autolysis (Cronshaw and Bouck, 1965; Wodzicki and Humphreys, 1972). Wodzicki and Humphreys (1972) suggest that the autolysis in tracheids is mediated by hydrolase-enriched vacuolar sap in a two step process: (1) structural changes in the membranes in contact with the vacuolar sap causes the formation of cytoplasmic spherules with a loss of continuity in the intravacuolar cytoplasmic filaments, and (2) an increase in membrane permeability allows hydrolytic enzymes to flood in from the vacuole and hydrolise the protoplast. Some of the hydrolytic enzymes may actually come from the wall itself (Olszewska *et al.*, 1966; Berlyn, 1979). In addition, the hydrolysed protoplast must be eluted out of the cells or plastered up against the cell wall (see Berlyn, 1964, 1970). These processes are also a part of tracheid differentiation.

What controls the duration of the cell cycle in the cambium and the rate and qualities of tracheid differentiation? The answers are unknown and yet they are of critical importance for wood quality and wood production. Conventional wisdom stresses the role of growth regulators produced in the leaves and transported to the cambium and developing xylem. However, the formation of these compounds requires time signals and once formed the growth regulators act as triggers that initiate morphogenetic processes. Once initiated the information system of the cell must activate, modulate, and sequence the biochemical systems of the cell.

The enlargement of the primary cell wall is intimately associated with hydroxyproline-rich glycoproteins. At present there are three somewhat conflicting hypotheses about the relationship between cell wall glycoproteins and cell enlargement (see Basile, 1980). The first can be called the Promotive Hypothesis and is due to Lamport (1963, 1965, 1967, 1969, 1970). Lamport termed these proteins 'extensins' implying that they promote cell extension. They have a polypeptide backbone with oligosaccharide side chains and could thus constitute a network of labile crosslinkages between cellulose microfibrils and perhaps other cell wall polysaccharides. Upon suitable stimulation such as auxin-dependent hydrogen ion excretion through the cell membrane, the crosslinkages break, cell wall synthesis begins, water uptake increases, and the cell enlarges.

An Inhibitive Hypothesis was developed by Cleland and Karlnes (1967) who proposed that hydroxyproline containing proteins (hyp-proteins) functioned by causing a cessation of cell elongation. They suggested that an increase in these hyp-proteins could be a factor in limiting cell enlargement. A number of subsequent investigations have supported this view while still other studies have suggested that the increase in hyp-protein is not as important as the change in its relationship to the other cell wall constituents.

Recently Basile (1979, 1980) has proposed a Suppression Hypothesis. It incorporates the inhibitory function of hyp-proteins from the Inhibitive Hypothesis and from the Promotive Hypothesis it postulates that the functional state of conformation of the hyp-protein is important in explaining its morphogenetic function. It does not require a specific relationship between the amount of hyp-protein and suppression of cell enlargement nor does it require a specific relationship between hyp-proteins and cellulose microfibrils. Basile has used analogs of hydroxyproline (which interfere with its normal metabolism) on peas and soybeans. In both species elongation was increased beyond the point where it would normally be suppressed and in both cases there was a marked decrease in the hydroxyproline content of wall bound proteins. Thus he concluded that certain hyp-proteins regulate morphogenesis by suppressing cell enlargement and/or division in highly localised populations of cells. The developing secondary xylem meets the latter criterion very well and it may be fruitful to look at the various species of hyp-proteins to see if they have morphogenetic roles in all of the phases of xylem cell differentiation. At the very least it is important to realise that there are probably compounds other than hormones that have a suppressive function in morphogenesis.

If hyp-proteins are involved with the cessation of elongation what initiates this phase? The answer appears to be indole-acetic-acid (IAA). Although there is opposition (see Terry and Jones, 1981) the currently most popular view is that IAA causes the release of protons into the cell wall (Cleland, 1971, 1973, 1976, 1977; Rayle and Cleland, 1977) and this 'loosens' the cell wall and causes a decrease in the cell's water potential (Stevenson and Cleland, 1981). Water enters the cell, turgor pressure increases, and the loosened wall stretches. At the same time cellulose synthesis occurs and restores the rigidity of the wall. At this point the process may be repeated or, if the proper morphogenetic conditions prevail, suppressed. The cell or a particular part of a cell may then be switched to the next phase of differentiation – wall thickening. However, it must be noted that the tips of the cell may be in the phase of elongation while more central regions are already in the phase of wall thickening and thus a single cell may span two different phases of differentiation. Also recent work suggests that cytokinins and gibberellins can also cause wall loosening but their relation to wall acidification is not known (Thomas *et al.*, 1981).

In most of the experimental systems investigated it it takes 10-15 minutes for auxin treatment to cause an increase in cell enlargement. This is too short a time for new protein synthesis to occur. Direct addition of protons also causes an increase in cell enlargement and here the lag time is only 1-2 minutes. This suggests either that the activation of an ATP-dependent proton pump by IAA takes 8-12 minutes, or that the protons are first packaged into vesicles which are then excreted through the plasmalemma into the cell wall. Vesicular excretion into the cell walls has been observed (Berlyn, 1970; Catesson, 1980a). The protons in the wall may function directly to loosen the wall by physical changes (Bates and Ray, 1981) or indirectly to loosen the wall by increasing the activity of hydrolytic enzymes. When a sufficient number of polysaccharide and/or pectic bonds are broken, the wall may become loosened and cell elongation can occur. This is a multistep process and new wall constituents (which may be contained in the vesicles) must continually be added in order for the system to operate. (Cell walls seldom become much thinner during the elongation phase of secondary xylem differentiation). Thus for long-term response to auxin the synthesis of both structural and catalytic proteins is required. In a quantitative cytochemical study of differentiating metaxylem vessels of *Zea mays* Patel and Berlyn (1981b) showed that as the stages of differentiation proceeded there was a progressive increase in RNA synthesis; this effect could be amplified by the addition of exogenous cytokinin. Other studies have reported that exogenous IAA increased RNA synthesis (Guilfoyle *et al.*, 1975). Since inhibitors of DNA synthesis do not block auxin effects on xylem differentiation exclusive of cell division, it has been suggested that auxin acts at the level of transcription or translation, for example, by increasing the activity of RNA polymerase (Guilfoyle *et al.*, 1975). Thus in some manner plant hormones do affect nunclear function but no specific mechanisms or nuclear receptor sites are known. Certainly the qualitative specificity demonstrated with the steroid hormones of animals is lacking and the functional mechanism in plants seems to be of a more quantitative nature.

One of the possible ways that the elongation phase is terminated may be by suppression of auxin. The plant cell has many mechanisms to get rid of IAA, *e.g.* binding IAA to peptides and esters, and oxidative degradation. Several IAA oxidase enzymes exist and all or nearly all are practically indistinguishable from the peroxidases which are involved in lignification. Possibly as the peroxidases are synthesised they first function as IAA oxidases to shift the xylem cell to the wall thickening phase and then as they increase in amount and/or activity they catalyse lignification.

It is important to consider where the hormone (information) signals come from that result in the sequential turning on of the metabolic systems that function in the six phases of wood formation and what causes the time signals *(Zeitgebers)* to be sent, *e.g.* quality and duration of solar radiation, mechanical or physiological stressors, metabolic feedbacks, *etc.*

Cambial stimulus

The cambial stimulus is one of the most intensively studied areas of forest biology and many relatively recent reviews are available (Wareing, 1958; Soding, 1961; Larson, 1962; Wilcox, 1962; Wort, 1962; Reinders-Gouwentak, 1965; Zimmermann and Brown, 1971, p. 75-85; Kramer and Kozlowski, 1979; Little and Wareing, 1981). Our discussion will be limited to certain morphogenetic questions and since the vast literature on the topic deals almost exclusively with trees of the temperate zone, our discussions will also be limited to this data base. Nevertheless generalised concepts of seasonal growth patterns in temperate trees must at least be compatible with the corresponding patterns in tropical trees. Radial growth in tropical tropical trees may be annual, semi-annual, irregular or continuous (Tomlinson and Longman, 1981). These patterns can vary between juvenile and mature phases in the same tree. In addition a single tree crown in the tropics may show non-synchronous behaviour with different branches flushing at different times of the year. Thus the dogma concerning the close correlation between short extension and radial growth that has emerged from the study of temperate trees may have to be modified in the tropics (see Bormann and Berlyn, 1981).

There are cases of anomalous growth that also require consideration. For example, in certain species, sectors of the vascular cambium become unidirectional, producing extensive amounts of phloem and little, if any, xylem (Dobbins, 1971, 1981). This results in sectors of phloem deeply embedded in the secondary xylem but also situated opposite the major vascular strands of the primary xylem. Such patterns suggest a morphogenetic correlation between the unidirectional cambial activity and the vascular pattern of the primary plant body. However, in the whorled shoots of *Clytostoma callistegioides,* the initial sectors form opposite the vascular strands, but as secondary growth continues additional sectors can form between *any* two existing sectors. The morphogenetic factors involved in this type of differentiation are not known.

According to Kramer and Kozlowski (1979) in a number of species (*e.g.* some members of Amaranthaceae, Avicenniaceae, Chenopodiaceae, Menispermaceae, Nyctaginaceae, *Gnetum, Welwitschia* and Cycadales) a series of successive bidirectional cambia are formed. In *Avicennia* these successive cambia result in the occurrence of alternating rings of xylem and phloem but the rings are not annual and in fact their number is correlated with the growth rate of the stem (Gill, 1971). No seasonality factors, if such exist, have been found that control this rhythmic pattern of cambial formation and activity. There are other woody species in the tropics that form interxylary phloem and the morphogenetic (and phylogenetic) factors involved in this are also unknown (Carlquist, 1981).

Early workers (*e.g.* Hartig, 1853; Jost, 1891, 1893; Priestly, 1930, 1932) postulated that buds and developing leaves produce some stimulus that travels downward from the shoot tips (basipetally) and reactivates the cambium in the spring. After the discovery of auxin many studies showed that exogenous auxin could initiate and stimulate cambial activity and auxin was extracted from the cambial region of a number of species. For reviews of the older literature see Soding (1961), Larson (1962), Wilcox (1962) and Reinders-Gouwentak (1965). These voluminous studies established that auxin can reactivate the cambium under experimental conditions and suggested that auxin is the agent that is involved in the *in vivo* process. However, a serious problem for the auxin hypothesis is the fact that the rate of polar auxin transport is not rapid enough to account for the rapidity of cambial reactivation.

Huber (1948) observed that bark slippage moved basipetally at the rate of *ca.* 42 mm hr^{-1}. This of course varies with species, tree height, tree vigour, crown dimensions and environmental factors like temperature. The currently accepted rate of basipetal (polar) auxin movement is 9-10 mm hr^{-1} (Hollis and Tepper, 1971; Goldsmith, 1977; Little, 1981) and thus there is almost a fivefold discrepancy between the rate of cambial reactivation and the rate of auxin movement. With the presently accepted rate of auxin movement it would take auxin over 200 days to travel 150 feet from the base of the crown of a large tree to stump height. This suggests that: (a) the currently accepted rate of auxin transport is too low; (b) there is a prior and more rapidly moving signal that activates the cambium and possibly stimulates the cambium and/or surrounding tissue to synthesise and/or release auxin and perhaps other growth regulators; or (c) cambial activation is not a function of the absolute amount of auxin. The possibilities for the fast-moving signal, if such exists, are unlimited but one plausible candidate would be electrical signals moving via cell membranes, especially the plasmalemma. Little (1981) has found that the rate of auxin transport is the same in resting and active cambia and thus he concluded that the auxin transport system and the structural changes associated with cambial reactivation are not closely linked. Most dividing and growing tissues produce auxin and the cambium and differentiating xylem are probably no exceptions (Mirov, 1941; Gunckel and Thimann, 1949; Titman and Wetmore, 1955; Soding, 1961; Sheldrake, 1971, 1973). Wareing (1951) proposed that rapid reactivation of cambium in ring-porous species is due to reserves of auxin precursor stored in the cambium. Wort (1962) suggested that tissues adjacent to the cambium can supply auxin. Thus auxin may be necessary for cambial activation but it need not come from the buds and it may not be the initial *'Zeitgeber'*.

Little and Wareing (1981) concluded that the cambial protoplast changes in its sensitivity to IAA such that the transition to true dormancy (rest) in the fall is the result of a morphogenetically induced inability to respond to auxin. As the dormancy changes to winter dormancy (quiescence), the cambial protoplasts regain the ability to respond to IAA. Their results did not substantiate a role for ABA (abscisic acid) in controlling cambial dormancy. However, water stress greatly curtailed foliar hormone production and thus indirectly affected cambial activity. Nevertheless, the intrinsic morphogenetic pattern controlling the cessation of cambial activity could not be circumvented by addition of exogenous IAA under natural field conditions or short days.

Morphogenetic domains

Hejnowicz (1961, 1964, 1968, 1971, 1973, 1975a, b) and his colleagues (Hejnowicz and Krawczyszyn, 1969; Krawczyszyn, 1972; Hejnowicz and Romberger, 1973) investigated the basis for the various types of grain pattern in trees (spiral left (S), spiral right (Z), interlocked, wavy) and in doing so have developed the concept that the cambium contains moving and pulsating domains. The oriented morphogenetic events that lead to grain variation are pseudotransverse divisions of fusiform initials, ray splitting and uniting, and apical intrusive growth of the fusiform initials. A cambial domain is defined as an area in the cambium where these morphogenetic events are non-random. Within a given domain the events are predominantly oriented in either S (left) or Z (right) direction. The position of these domains on the cambial surface is termed the domain pattern. However, the orientation of the cambium is not necessarily reflected in the orientation of the cells of the xylem. In general the correlation between cambial orientation and wood element orientation increases as the rates of the morphogenetic events increase. The domains may move up or down and bidirectional movement can occur in a single tree stem. Furthermore, in addition to migrating, the domains can grow and shrink in size (pulsate) and even disappear and reform. A single cambial initial may switch from S to Z orientation. The origins of these domains are unknown but possibilities include internal clocks and maps, morphogenetic waves and their superposition, chemical gradients, and hormone flow from the leaves. Hejnowicz believes that the origins of the domains are morphogenetic waves of different wave-length and velocity but constant period, T. The superposition of these variable length but period invariant waves produces a stationary pattern of amplitudes that serves as a morphogenetic map. The propagation of waves within this pattern is controlled by a biological clock. The origin of the waves is a hypothetical set of period-invariant oscillators in the protoplasm.

These oscillators are programmed by one set of genes and another set of genes produces factors that are responsible for the velocity of signalling between the oscillators (Hejnowicz, 1974).

Hejnowicz (1975b) concludes that the cambium is subject to two rhythms: one that controls the oriented morphogenetic events and that can result in a specific pattern of grain inclination or conformation, and a second externally driven rhythm that results in wood production and is manifested as 'rings' of wood density. However, these two rhythms are not independent since the frequency of the oriented morphogenetic events is not independent of radial growth rate although the relationship between the two is quite complex. The orientation rhythm is also tied to calendar time since pseudotransverse divisions tend to be more frequent late in the growing season. Also, as mentioned previously, it is necessary to consider wood formation as a three dimensional process and a process that is integrated into the overall pattern of stem form pattern in the tree.

The concept of morphogenetic domains has recently been extended by the incorporation of schemes for hormone modulation of the system (Zajaczkowski and Wodzicki, 1978; Wodzicki *et al.*, 1979). According to these workers auxin is produced by the stem segments in pulses which result in a wave-like pattern. The cambial zone and adjacent tissues are said to be composed of polar cells that generate oscillations. The wavelength of the auxin waves is several times longer than the cell length in the cambial region and hence the oscillatory system is thought to be supracellular. The auxin wave disappears if both IAA and ABA are simultaneously applied to the top of the stem segment. This modulating effect travels faster than the known rate of auxin transport and this is said to be a manifestation of apical control of the conjugated oscillators that establish the morphogenetic field regulating cambial activity. However, the reality of this system is yet to be established. The methodology involves taking stem segments and applying agar blocks (some with and some without growth regulators). Then the stem segments are cut into smaller segments and auxin is allowed to diffuse out into agar strips and the amount quantified. According to Zajaczkowski (seminar, 1981), the segments require a 100 minute incubation period before oscillators are observed and smaller segments transport more auxin than larger ones. This suggests that the cutting could induce the oscillations. The oscillations could for example be a boundary or wound effect. Another question is whether there is more total IAA produced or whether there is a change in the ratio of bound to free auxin, possibly induced by the experimental procedures.

Biophysical factors

One of the main biological functions of wood in trees is to provide structural support for the entire tree and to project the foliar organs into receptive positions with respect to light. The wood of roots also has mechanical capacities in its role of anchorage and substrate penetration. Thus mechanical stresses and perhaps even potential stresses play a role in wood morphogenesis. Furthermore, the strength of a tree stem increases with the amount of sub-inhibitory stress they are exposed to. In fact the strength of plant parts such as petioles can be greatly (10 fold) increased by artificial weight-training programs (Newcombe, 1895). In addition, normal wood formation by the cambium requires a certain amount of pressure (Brown and Sax, 1962; Brown, 1964). Originally the cause of reaction wood was thought to be due to mechanical stress and some workers (especially those with engineering backgrounds) still espouse this theory (*cf.* Boyd, 1977; Wilson and Archer, 1977). Although some research continues in this area (*e.g.* Kellogg and Steucek, 1977) it has been largely neglected and certainly deserves more consideration.

References

Bailey, I.W. 1919. Phenomena of cell division in the cambium of arborescent gymnosperms and their cytological significance. Proc. Nat. Acad. Sci. 5: 283-285.

Bailey, I.W. 1920a. The formation of the cell plate in the cambium of the higher plants. Proc. Nat. Acad. Sci. 6: 197-200.

Bailey, I.W. 1920b. The significance of the cambium in the study of certain physiological problems. J. Gen. Physiol. 2: 519-533.

Bailey, I.W. 1920c. The cambium and its derivative tissues. II. Size variations of cambial initials in gymnosperms and angiosperms. Amer. J. Bot. 7: 355-367.

Bailey, I.W. 1920d. The cambium and its derivative tissues. III. A reconnaissance of cytological phenomena in the cambium. Amer. J. Bot. 7: 417-434.

Bailey, I.W. 1923. The cambium and its derivative tissues. IV. The increase in girth of the cambium. Amer. J. Bot. 10: 499-509.

Bailey, I.W. 1930. The cambium and its derivative tissues. V. A reconnaissance of the vacuome in living cells. Zeitschr. Mikr. Anat. 10: 651-682.

Bailey, I.W. & T. Kerr. 1934. The cambium and its derivative tissues. X. Structure, optical properties and chemical composition of the so-called middle lamella. J. Arn. Arbor. 16: 273-300.

Bannan, M.W. 1950. The frequency of anticlinal divisions in fusiform cambial cells of Chamaecyparis. Amer. J. Bot. 37: 511-519.

Bannan, M.W. 1951. The annual cycle of size changes in the fusiform cambial cells of Chamaecyparis and Thuja. Canad. J. Bot. 29: 421-437.

Bannan, M.W. 1953. Further observations on the reduction of fusiform cambial cells in Thuja occidentalis L. Canad. J. Bot. 31: 63-74.

Bannan, M.W. 1956. Some aspects of the elongation of fusiform cambial cells in Thuja occidentalis L. Canad. J. Bot. 34: 175-196.

Bannan, M.W. 1957a. Girth increase in white cedar stems of irregular form. Canad. J. Bot. 35: 425-434.

Bannan, M.W. 1957b. The relative frequency of the different types of anticlinal divisions in conifer cambium. Canad. J. Bot. 35: 875-884.

Bannan, M.W. 1962. Cambial behavior with reference to cell length and ring width in Pinus strobus L. Canad. J. Bot. 40: 1057-1062.

Bannan, M.W. 1967. Anticlinal divisions and cell length in conifer cambium. For. Prod. J. 17: 63-69.

Bannan, M.W. 1968a. Anticlinal divisions and the organization of conifer cambium. Bot. Gaz. 129: 107-113.

Bannan, M.W. 1968b. Polarity in the survival and elongation of fusiform initials in conifer cambium. Canad. J. Bot. 46: 1005-1008.

Bannan, M.W. 1970. A survey of cell length and frequency of multiplicative divisions in the cambium of conifers. Canad. J. Bot. 48: 1585-1589.

Bannan, M.W. & M. Bindra. 1970. Variations in cell length and frequency of anticlinal division in the vascular cambium throughout a white spruce tree. Canad. J. Bot. 48: 1363-1371.

Barghoorn, E.S. 1964. Evolution of cambium in geologic time. In: M.H. Zimmermann (ed.), The Formation of Wood in Forest Trees: 3-18. Acad. Press, New York.

Barnett, J.R. 1973. Seasonal variation in the ultrastructure of the cambium in New Zealand grown Pinus radiata D. Don. Ann. Bot. 37: 1005-1011.

Barnett, J.R. 1975. Seasonal variation of organelle members in sections of fusiform cambium cells of Pinus radiata D. Don. New Zeal. J. Bot. 13: 325-332.

Basile, D.V. 1979. Hydroxyproline-induced changes in form, apical development and cell wall protein in the liverwort, Plagiochila arctica. Amer. J. Bot. 66: 776-783.

Basile, D.V. 1980. A possible mode of action for morphoregulatory hydroxyproline-proteins. Bull. Torrey Bot. Club. 107: 325-338.

Bates, G.W. & P.M. Ray. 1981. pH-dependent interactions between pea cell wall polymers possibly involved in wall deposition and growth. Plant Physiol. 68: 158-164.

Berlyn, G.P. 1961. Factors affecting the incidence of reaction tissue in Populus deltoides Bartr. Iowa State J. Sci. 35: 367-424.

Berlyn, G.P. 1962. Some size shape relationships between tree stems and crowns. Iowa State J. Sci. 37: 7-15.

Berlyn, G.P. 1963. Methacrylate as an embedding medium for woody tissues. Stain Technol. 38: 23-28.

Berlyn, G.P. 1964. Recent advances in wood anatomy: The cell walls in secondary xylem. For. Prod. J. 14: 467-476.

Berlyn, G.P. 1970. Ultrastructural and molecular concepts of cell-wall formation. Wood and Fiber 2: 196-227.

Berlyn, G.P. 1972. Seed germination and morphogenesis. In: T.T. Kozlowski (ed.), Seed Biology, Vol. 1: 223-312. Acad. Press, New York.

Berlyn, G.P. 1979. Physiological control of differentiation of xylem elements. Wood and Fiber 11: 109-126.

Berlyn, G.P. & R.C. Beck. 1980. Tissue culture as a technique for studying meristematic activity. In: C.H.A. Little (ed.), Control of Shoot Growth in Trees: 305-324. IUFRO Workshop Proc. Marit. For. Res. Center, Fredericton, New Brunswick, Canada.

Berlyn, G.P., S.S. Dhillon & J.P. Miksche. 1979. Feulgen cytophotometry of pine nuclei. II. Effect of pectinase used in cell separation. Stain Technol. 54: 201-204.

Berlyn, G.P. & J.P. Miksche. 1976. Botanical microtechnique and cytochemistry. Iowa State Univ. Press, Ames, Iowa.

Bormann, F.H. & G.P. Berlyn (eds.). 1981. Age and growth rate of tropical trees. Yale Univ. School For. Environm. Studies, Bull. No. 94. New Haven, Connecticut.

Boyd, J.D. 1977. Basic cause of differentiation of tension wood and compression wood. Austr. For. Res. 7: 121-143.

Brown, C.L. 1964. The influence of external pressure on the differentiation of cells and tissues cultured in vitro. In: M.H. Zimmermann (ed.), The Formation of Wood in Forest Trees: 389-404. Acad. Press, New York.

Brown, C.L. & K. Sax. 1962. The influence of pressure on the differentiation of secondary tissues. Amer. J. Bot. 49: 683-691.

Carlquist, S. 1981. Types of cambial activity and wood anatomy of Stylidium (Stylidiaceae). Amer. J. Bot. 68: 778-785.

Catesson, A.M. 1964. Origine, fonctionnement et variations cytologiques saisonnieres du cambium de l'Acer pseudoplatanus L. (Acéracées). Ann. Sc. Nat. Bot. 12e sér. 5: 229-498.

Catesson, A.M. 1980a. The vascular cambium. In: C.H.A. Little (ed.), Control of Shoot Growth in Trees: 12-40. IUFRO Workshop Proc. Marit. For. Res. Center, Fredericton, New Brunswick, Canada.

Catesson, A.M. 1980b. Le cycle saisonnier des cellules cambiales chez quelques Feuillus. Actualités Botaniques (not seen, quoted in Catesson 1980a op. cit.).

Catesson, A.M. & J.C. Roland. 1981. Sequential changes associated with cell wall formation and fusion in the vascular cambium. IAWA Bull. n.s. 2: 151-162.

Cleland, R. 1971. Cell wall extension. Ann. Rev. Plant Physiol. 22: 197-220.

Cleland, R. 1973. Auxin-induced hydrogen ion excretion from Avena coleoptiles. Proc. Nat. Acad. Sci. 70: 3092-3093.

Cleland, R. 1976. Kinetics of hormone-induced H$^+$ excretion. Plant Physiol. 58: 210-213.

Cleland, R. 1977. The control of cell enlargement. In: D.H. Jennings (ed.), Integration of Activity in the Higher Plant. Soc. Exp. Biol. Symp. 31: 101-115.

Cleland, R. & A. Karlnes. 1967. A possible role of hydroxyproline-containing proteins in the cessation of cell elongation. Plant Physiol. 42: 669-671.

Cottignies, A. 1979. The blockage in the G1 phase of the cell cycle in the dormant shoot apex of ash. Planta 147: 15-19.

Cronshaw, J. 1965. The formation of the wart structure in tracheids of Pinus radiata. Protoplasma 60: 233-242.

Cronshaw, J. & J.B. Bouck. 1965. The fine structure of differentiating xylem elements. J. Cell Biol. 24: 415-431.

Dadswell, H.W. 1963. Tree growth-wood property inter-relationships. I. Need for knowledge of effects of growth conditions on cell structure and wood properties. In: T.E. Maki (ed.), Proc. special Field Institute in Forest Biology: 5-11. School For., North Carolina State Univ., Raleigh, N.C.

Dhillon, S.S., G.P. Berlyn & J.P. Miksche. 1978. Nuclear DNA content in populations of Pinus rigida. Amer. J. Bot. 65: 192-196.

Dobbins, D.R. 1971. Studies on anomalous cambial activity in the Bignoniaceae. II. A case of differential production of vascular tissues. Amer. J. Bot. 49: 2107-2110.

Dobbins, D.R. 1981. Anomalous secondary growth in lianas of the Bignoniaceae is correlated with the vascular pattern. Amer. J. Bot. 68: 142-144.

Esau, K. 1948. Phloem structure in the grapevine and its seasonal changes. Hilgardia 18: 217-296.

Esau, K. 1977. Anatomy of seed plants. John Wiley & Sons, New York.

146

Evert, R.F. 1961. Some aspects of cambial development in Pyrus communis. Amer. J. Bot. 48: 479-488.

Farooqui, P. & A.W. Robards. 1979. Seasonal changes in the ultrastructure of cambium of Fagus silvatica L. Proc. Ind. Acad. Sci. B 88: 463-472.

Ford, E.D. & A.W. Robards. 1976. Short term variation in tracheid development in the early wood of Picea sitchensis. In: P. Baas, A.J. Bolton & D.M. Catling (eds.), Wood Structure in Biological and Technological Research: 212-221. Leiden Bot. Series No. 3. Leiden Univ. Press, The Hague.

Fosket, D.E. 1968. Cell division and the differentiation of wound-vessel members in cultured stem segments of Coleus. Proc. Nat. Acad. Sci. (USA) 59: 1089-1096.

Fosket, D.E. 1970. The time course of xylem differentiation and its relation to deoxyribonucleic acid synthesis in cultured Coleus stem segments. Plant Physiol. 46: 64-68.

Fosket, D.E. 1972. Meristematic activity in relation to wound xylem differentiation. In: M.H. Miller & C.C. Kuehnert (eds.), The Dynamics of Meristem Cell Populations: 33-50. Plenam Press, New York.

Fosket, D.E. & J.G. Torrey. 1969. Hormonal control of cell proliferation and xylem differentiation in cultured tissues of Glycine max var. Bioloxi. Plant Physiol. 44: 871-880.

Ghouse, A.K.M. & M. Yunus. 1974. The ratio of ray and fusiform initials in some woody species of the Ranalian complex. Bull. Torrey Bot. Club 101: 363-366.

Ghouse, A.K.M. & M. Yunus. 1976. Ratio of ray and fusiform initials in the vascular cambium of certain leguminous trees. Flora 165: 23-28.

Gill, A.M. 1971. Endogenous control of growth-ring development in Avicennia. For. Sci. 17: 462-465.

Goldsmith, M.H.M. 1977. The polar transport of auxin. Ann. Rev. Plant Physiol 28: 439-478.

Guilfoyle, T.J., C.Y. Lin, Y.M. Chen, R.T. Nagao & J.L. Key. 1975. Enhancement of soybean RNA polymerase I by auxin. Proc. Nat. Acad. Sci. (USA) 72: 69-72.

Gunckel, J.E. & K.V. Thimann. 1949. Studies of development in long shoots and short shoots of Ginkgo biloba. Amer. J. Bot. 36: 145-151.

Hartig, T. 1853. Über die Entwickelung des Jahrringes der Holzpflanzen. Bot. Zeitung 11: 553-556, 569-579.

Hartig, R. 1901. Holzuntersuchungen, Altes und Neues. Julius Springer, Berlin.

Hejnowicz, Z. 1961. Anticlinal division, intrusive growth, and loss of fusiform initials in non-storied cambium. Acta Soc. Bot. Pol. 30: 729-748.

Hejnowicz, Z. 1964. Orientation of the partition in pseudotransverse division in cambia of some conifers. Canad. J. Bot. 42: 1685-1691.

Hejnowicz, Z. 1968. The structural mechanism involved in the changes of grain in timber. Acta Soc. Bot. Pol. 37: 347-365.

Hejnowicz, Z. 1971. Upward movement of the domain pattern in the cambium producing wavy grain in Picea excelsa. Acta Soc. Bot. Pol. 40: 499-512.

Hejnowicz, Z. 1973. Morphogenetic waves in cambia of trees. Plant Sci. Letters 1: 359-366.

Hejnowicz, Z. 1974. Study of migrating orientational domain patterns in cambia of trees as a morphogenic wave phenomenon and its manifestation in wood grain. Report No. 2. Bot. Inst. Univ. Wroclaw, Poland.

Hejnowicz, Z. 1975a. Study of migrating orientational domain pattern in cambia of trees as a morphogenetic wave phenomenon and its manifestation in wood grain. Report No. 4. Bot. Inst. Univ. Wroclaw, Poland.

Hejnowicz, Z. 1975b. A model for morphogenetic map and clock. J. Theor. Biol. 54: 345-362.

Hejnowicz, A. & J. Krawczyszyn. 1969. Oriented morphogenetic phenomena in cambium of broad leaved trees. Acta Soc. Bot. Pol. 38: 547-560.

Hejnowicz, Z. & J.A. Romberger. 1973. Migrating cambial domains and the origin of wavy grain in xylem of broad-leaved trees. Amer. J. Bot. 60: 209-222.

Hollis, C.A. & H.B. Tepper. 1971. Auxin transport within intact dormant and active white ash shoots. Plant Physiol. 48: 146-149.

Huber, B. 1948. Physiologie der Rindenschalung bei Fichte und Eichen. Forstwiss. Centralbl. 67: 129-164.

Isebrands, J.G. & P.R. Larson. 1973. Some observations on the cambial zone in cotton wood. IAWA Bull. 1973/3: 3-9.

Itoh, T. 1971. On the ultrastructure of dormant and active cambium of conifers. Wood Res. 51: 33-45.

Jost, L. 1891. Über Dickenwachstum und Jahresringbildung. Bot. Zeitung 49: 482-499.

Jost, L. 1893. Über die Beziehungen zwischen der Blattentwickelung und der Gefässbildung in der Pflanze. Bot. Zeitung 51: 89-138.

Kellogg, R.M. & G.L. Steucek. 1977. Motion-induced growth effects in Douglas fir. Canad. J. For. Res. 7: 94-99.

Kennedy, R.W. & J.L. Farrar. 1965. Tracheid development in tilted seedlings. In: W.A. Côté (ed.), Cellular Ultrastructure of Woody Plants: 419-453. Syracuse Univ. Press, Syracuse.

Kramer, P.J. & T.T. Kozlowski. 1979. Physiology of woody plants. Acad. Press, New York.

Krawczyszyn, J. 1972. Movement of the cambial domain pattern and mechanism of formation of interlocked grain in Platanus. Acta Soc. Bot. Pol. 41: 443-461.

Lamport, D.T.A. 1963. O_2 fixation into hydroxyproline of plant cell wall protein. J. Biol. Chem. 238: 1438-1440.

Lamport, D.T.A. 1965. The protein component of primary cell walls. Adv. Bot. Res. 2: 151-218.

Lamport, D.T.A. 1967. Hydroxyproline-0-glycosidic linkage of the plant glycoprotein extension. Nature 216: 1322-1324.

Lamport, D.T.A. 1969. The isolation and partial characterization of hydroxyproline-rich glycopeptides obtained by enzymic degradation of primary cell walls. Biochem. 8: 1155-1163.

Lamport, D.T.A. 1970. Cell wall metabolism. Ann. Rev. Plant Physiol. 21: 235-270.

Larson, P.R. 1962. Auxin gradients and the regulation of cambial activity. In: T.T. Kozlowski (ed.), Tree Growth: 97-117. Ronald Press, New York.

Little, C.H.A. 1981. Effect of cambial dormancy state on the transport of [$1^{-14}C$] indol-3-ylacetic acid in Abies balsamea shoots. Canad. J. Bot. 59: 342-348.

Little, C.H.A. & P.F. Wareing. 1981. Control of cambial activity and dormancy in Picea sitchensis by indol-3-ylacetic and abscisic acids. Canad. J. Bot. 59: 1480-1492.

Mahmood, A. 1968. Cell groupings and primary wall generation in the cambial zone, xylem, and phloem in Pinus. Austr. J. Bot. 16: 177-195.

Mahmood, A. 1971. Numbers of initial-cell divisions as a measure of activity in the yearly cambial growth pattern in Pinus. Pakistan J. For. 21: 27-42.

Mia, A.J. 1970. Fine structure of active, dormant, and aging cambial cells in Tilia americana. Wood Sci. 3: 34-42.

Mirov, N.T. 1941. Distribution of growth hormone on shoots of two species of pine. J. Forestry 39: 457-464.

Murmanis, L. 1970. Locating the initial in the vascular cambium of Pinus strobus by electron microscopy. Wood Sci. Technol. 4: 1-17.

Murmanis, L. 1971. Structural changes in the vascular cambium of Pinus stribus L. during an annual cycle. Ann. Bot. 35: 133-141.

Newcombe, F.C. 1895. The regulatory formation of mechanical tissue. Bot. Gaz. 20: 441-448.

Newman, I.V. 1956. Pattern in meristems of vascular plants. I. Cell partition in living apices and in the cambial zone in relation to the concepts of initial cells and apical cells. Phytomorphology 6: 1-19.

148

Oliver, C.D. 1978. The development of northern red oak in mixed stands in central New England. Yale Univ. School For. Environm. Studies, Bull. No. 91. New Haven, Connecticut.

Oliver, C.D. 1980. Even-aged development of mixed-species stands. J. Forestry 78: 201-203.

Olszewska, M.J., B. Gabara & F. Steplewski. 1966. Recherches cytochimiques sur la succession d'enzymes hydrolytiques, sur la présence de la thiamine pyrophosphatase et des polysaccharides au cours du développement de la plague cellulaire. Protoplasma 61: 60-80.

Owens, J.N. & M. Molder. 1973. A study of DNA and mitotic activity in the vegetative apex of Douglas fir during the annual growth cycle. Canad. J. Bot. 51: 1395-1409.

Paliwal, G.S. & L.M. Srivastava. 1969. The cambium of Alseuosmia. Phytomorphology 19: 5-8.

Parthasarathy, M.V. & T.C. Pesacreta. 1980. Microfilaments in plant vascular cells. Canad. J. Bot. 58: 807-815.

Patel, K.R. & G.P. Berlyn. 1981a. Genetic instability of multiple buds of Pinus coulteri regenerated from tissue culture. In press.

Patel, K.R. & G.P. Berlyn. 1981b. Influence of kinetin on histone composition and endogenous RNA level in differentiating metaxylem of Zea mays root tips. Unpublished manuscript.

Priestly, J.H. 1930. Studies in the physiology of cambial activity. III. The seasonal activity of the cambium. New Phytol. 29: 316-354.

Priestly, J.H. 1932. The growing tree. Forestry 6: 105-112.

Rayle, D. & R. Cleland. 1977. Control of plant cell enlargement by hydrogen ions. Developmental Biology 11: 187-211.

Reinders-Gouwentak, C.A. 1965. Physiology of the cambium and other secondary meristems of the shoot. In: W. Ruhland (ed.), Encyclopedia of Plant Physiology Vol. 15: 1076-1105. Springer, Berlin.

Robards, A.W. & P. Kidway. 1969. A comparative study of the ultrastructure of resting and active cambium of Salix fragilis L. Planta 84: 239-249.

Roberts, L.W. 1969. The initiation of xylem differentiation. Bot. Rev. 35: 201-250.

Roberts, L.W. 1976. Cytodifferentiation in plants. Cambridge Univ. Press, New York.

Roland, J.C. 1978. Early differences between radial walls and tangential walls of actively growing cambial zone. IAWA Bull. 1978/1: 7-10.

Sanio, K. 1873. Anatomie der gemeinen Kiefer (Pinus sylvestris L.). Jahrb. Wiss. Bot. 9: 50-126.

Sheldrake, A.R. 1971. Auxin in the cambium and its differentiating derivatives. J. Exp. Bot. 22: 735-740.

Sheldrake, A.R. 1973. The production of hormones in higher plants. Biol. Rev. 48: 509-559.

Sinnott, E.W. 1960. Plant morphogenesis. McGraw-Hill, New York.

Skene, D.S. 1972. The kinetics of tracheid development in Tsuga canadensis Corr. and its relation to tree vigour. Ann. Bot. 36: 179-187.

Soding, H. 1961. Vorkommen und Verteilung der Auxine der Pflanze. In: W. Ruhland (ed.), Encyclopedia of Plant Physiology: 583-699. Springer, Berlin.

Srivastava, L.M. 1966. On the fine structure of Fraxinus americana. L. J. Cell Biol. 31: 79-93.

Srivastava, L.M. & T.P. O'Brien. 1966. On the ultrastructure of cambium and its vascular derivatives. I. Cambium of Pinus strobus. Protoplasma 61: 257-276.

Stevenson, T.T. & R. Cleland. 1981. Osmoregulation in the Avena coleoptile in relation to auxin and growth. Plant Physiol. 67: 749-753.

Stewart, C.M. 1957. Status of cambial chemistry. TAPPI 40: 244-256.

Stewart, C.M. 1966. The chemistry of secondary growth in trees. Div. For. Prod. Techn. Paper No. 43. CSIRO, Melbourne.

Terry, M.E. & R.L. Jones. 1981. Effect of salt on auxin-included acidification and growth by pea internode sections. Plant Physiol. 68: 59-64.

Thomas, J. & C.W. Ross, C.J. Chastain, N. Koomanoff, J. Hendrix & E. Van Volkenburgh. 1981. Cytokinin-induced wall extensibility in excised cotyledons of radish and cucumber. Plant Physiol. 68: 107-110.

Timell, T.E. 1973. Ultrastructure of the dormant and active cambial zones and the dormant phloem associated with formation of normal and compression woods in Picea abies (L.) Karst. SUNY Coll. Environm. Sci. For. Techn. Publ. No. 96. Syracuse, New York.

Timell, T. 1979. Formation of compression wood in balsam fir (Abies balsamea). I. Ultrastructure of the active cambial zone and its enlarging derivatives. Holzforschung 33: 137-143.

Timell, T.E. 1980a. Organization and ultrastructure of the dormant cambial zone in compression wood of Picea abies. Wood Sci. Technol. 14: 161-179.

Timell, T.E. 1980b. Karl Gustav Sanio and the first scientific description of compression wood. IAWA Bull. n.s. 1: 147-153.

Titman, P.W. & R.H. Wetmore. 1955. The growth of long and short shoots of Cercidiphyllum. Amer. J. Bot. 42: 364-372.

Tomlinson, P.B. & K.A. Longman. 1981. Growth phenology of tropical trees in relation to cambial activity. In: F.H. Bormann & G.P. Berlyn (eds.), Age and Growth Rate of Tropical Trees: 7-19. Yale Univ. School For. Environm. Studies, Bull. No. 94. New Haven, Connecticut.

Tsuda, M. 1975. The ultrastructure of the vascular cambium and its derivatives in coniferous species. I. Cambial cells. Bull. Tokyo Univ. For. 67: 158-226.

Wardrop, A.B. 1952. Formation of new cell walls in cell division. Nature 170: 329.

Wardrop, A.B. 1957. The phase of lignification in the differentiation of wood fibres. TAPPI 40: 225-243.

Wardrop, A.B. 1964. The reaction anatomy of arborescent angiosperms. In: M.H. Zimmermann (ed.), The Formation of Wood in Forest Trees: 405-456. Acad. Press, New York.

Wareing, P.F. 1951. Growth studies in woody species. IV. The initiation of cambial activity in ring-porous species. Physiol. Plant 4: 546-562.

Wareing, P.F. 1958. The physiology of cambial activity. J. Inst. Wood Sci. 1: 34-42.

Westing, A.H. 1965. The formation and function of compression wood in gymnosperms. I. Bot. Rev. 31: 381-480.

Westing, A.H. 1968. The formation and function of compression wood in gymnosperms. II. Bot. Rev. 34: 51-105.

Whalley, B.E. 1950. Increase in girth of the cambium in Thuja occidentalis L. Canad. J. Res. C 28: 331-340.

Wilcox, H. 1962. Cambial growth characteristic. In: T.T. Kozlowski (ed.), Tree Growth: 57-88. Ronald Press, New York.

Wilson, B.F. 1963. Increase in cell wall surface area during enlargement of cambial derivatives in Abies concolor. Amer. J. Bot. 50: 95-102.

Wilson, B.F. 1964. A model for cell production by the cambium of conifers. In: M.H. Zimmermann (ed.), The Formation of Wood in Forest Trees: 19-36. Acad. Press, New York.

Wilson, B.F. 1970. The growing tree. Univ. Massachusetts Press, Amherst.

Wilson, B.F. & R.R. Archer. 1977. Reaction wood: Induction and mechanical action. Ann. Rev. Plant Physiol 28: 23-43.

Wilson, B.F. & R.A. Howard. 1968. A computer model for cambial activity. For. Sci. 14: 77-90.

Wodzicki, T.J. & C.L. Brown. 1973. Organization and breakdown of the protoplast during maturation of pine tracheids. Amer. J. Bot. 60: 631-640.

Wodzicki, T.J. & W.J. Humphreys. 1972. Cytodifferentiation of maturing pine tracheids. Tissue and Cell 4: 525-528.

Wodzicki, T.J., A.B. Wodzicki & S. Zajaczkowski. 1979. Hormone modulation of the oscillatory system involved in polar transport of auxin. Physiol. Plant 46: 97-100.

Worrall, J.F. 1980. The impact of environment on cambial growth. In: C.H.A. Little (ed.), Control of Shoot Growth in Trees: 127-142. IUFRO Workshop Proc. Marit. For. Res. Center, Fredericton, New Brunswick, Canada.

Wort, D.J. 1962. Physiology of cambial activity. In: T.T. Kozlowski (ed.), Tree Growth: 89-95. Ronald Press, New York.

Zajaczkowski, S. & T.J. Wodzicki. 1978. Auxin and plant morphogenesis - a model of regulation. Acta Soc. Bot. Pol. 47: 233-243.

Zimmermann, M.H. (ed.). 1964. The formation of wood in forest trees. Acad. Press, New York.

Zimmermann, M.H. & C.L. Brown. 1971. Trees: structure and function. Springer, New York.

Zimmermann, M.H. & P.B. Tomlinson. 1966. Analysis of complex vascular systems in plants: Optical shuttle method. Science 15: 72-73.

Zimmermann, M.H. & P.B. Tomlinson. 1967. A method for the analysis of the course of vessels in wood. IAWA Bull. 1967/1: 2-6.

Genetic variation in wood properties

J. BURLEY

Department of Agricultural and Forest Sciences, Oxford University, Commonwealth Forestry Institute, South Parks Road, Oxford, England.

Summary: The significance of genetic variation in commercially important technological characters of wood has been realised only in the last 50 years: most of the initial studies were made by silviculturists and later by tree breeders, occasionally in cooperation with wood technologists rather than anatomists. The wood properties examined were not always those considered by taxonomic anatomists; these anatomists recognised the existence of variation in cell and tissue size, composition and distribution but accounted for it only to the extent of estimating mean values mainly at the specific level. The fundamental sources of variation (mutation, migration and selection) apply as much to anatomy as to morphology and many anatomical features are under polygenic control (metric, quantitative traits). Differences between species, populations and trees have been demonstrated for a wide range of wood properties in species, provenance and progeny trials of many species; trees also vary genetically in the extent and pattern of intra-tree variation. The type of genetic control (additive or non-additive) varies between traits. For precise estimation of population parameters many samples are required and modern techniques such as densitometry and image-analysis, coupled with computing and statistical methods, facilitate rapid processing and evaluation of large numbers for several wood traits. For many species, characters, and plantation sites, attention is now being concentrated on estimating genetic parameters including correlations between anatomical traits and utilisation characteristics.

Introduction

'It is true of all trees anywhere that
with a north aspect the wood is closer
and more compact and better generally.'
Theophrastus, 300 BC.

Despite the sweeping generalisation of this statement, made some 2,230 years ago in the northern hemisphere, Theophrastus may be credited with the first recorded observation that wood characteristics vary. The cause of the variation may have been purely environmental or a genetic adaptation to environmental differences but it was not until 2000 years later in the 18th century, that any indication of a major genetic effect was given; Hamilton (1761) observed differences in wood-using characteristics of Scots pine grown in Scotland from different seed origins. No practical use was made of such genetic differences until serious provenance testing began in the 20th century and only in the last 50 years have tree breeders begun to capitalise on the great variation between individual trees. Champion (1925, 1933) began a series of investigations into spiral grain of *Pinus roxburghii* that heralded an era of both controversy and progress in determining genetic control of anatomical features of wood.

Schreiner (1935) considered that strains of trees with special wood properties could be bred while Koehler (1939) felt that wood characteristics are conservative and more easily modified environmentally than genetically; however, he recognised the need to determine the genetic contribution in order to decide between genetic and silvicultural methods of improvement. Since that pre-war period and indeed since 1900, considerable research has been undertaken on individual wood properties, on the relations between anatomical, chemical, mechanical and physical properties and end use characteristics, and on the patterns of inheritance of many of these traits. Although most of the work on strength characteristics and end use performance has been done by timber technologists *sensu strictu,* much of the research on the variability of wood anatomy and morphology within and between trees has been carried out by or for silviculturists, forest geneticists and tree breeders. This is largely because many silviculturists and geneticists have accepted the improvement of wood quality as one of the main objectives of management and breeding.

For many political, social and phytogeographical reasons most of this effort has been concentrated in temperate regions. With the current rapid increase in plantation programmes, particularly with conifers and fast growing hardwoods, in tropical and developing countries, there is an urgent need for similar information in these areas. Further, with the rapid decline of tropical forests, particularly evergreen forests and savannah woodlands (Spears, 1979), there is an urgent need to undertake similar studies in species that may have potential for plantations. In addition to the application of wood anatomy to technology, anatomical features may be of use in taxonomy, identification, dendrochronology and even forest management* (Bamber, 1978a; Brazier, 1968). There is

---

* A recent workshop on determination of age and growth rate of tropical trees (Bormann and Berlyn, 1981) reviewed the application of various techniques and properties including wood anatomical features (Fahn *et al.*, 1981) and established an international working group (Co-chairmen J. Burley and G.P. Berlyn).

thus a need for classical wood anatomists to be aware of the extent of genetic variation in their chosen species. Traditional taxonomic descriptions of the wood anatomy of families, genera and species have often been based on limited sampling (small numbers of trees and small numbers or sizes of samples within trees); statistical analyses have commonly been ignored or were inappropriate (see the paper by Burley and Miller, 1982, elsewhere in this volume) so that mean values, presented with or without estimates of precision, do not reflect the total variability of a taxon nor the partitioning of the variability into genetic and non-genetic sources.

The existence of phenotypic variation in wood anatomy between populations of the same species as well as between species in a genus or family has been explained by Carlquist (1975) in terms of ecological adaptation to varying environmental conditions and subsequent evolution. Supporting evidence has been put forward in the form of altitudinal, latitudinal and xeric trends in hardwood vessel member length and in number of bars per scalariform perforation (Baas, 1973, 1976; Dickison *et al.*, 1978 Van der Graaff and Baas, 1974; Miller, 1976; Novruzova, 1968; Van den Oever *et al.*, 1980, 1981; Den Outer and Van Veenendaal, 1976) but, as Baas (1976) pointed out, functional interpretations of such variation would be far-fetched and impossible in most cases where specimens from natural habitats are involved (see Heywood, 1973). However, he recognised that genetic control of vessel member characters is likely and that natural selection could be expected to canalise the variation yielding anatomical correlations with ecological factors.

The only reliable way to estimate the degree and pattern of genetic variation within a species is by seed source experiments in which samples of seeds from each population are sown and the resultant plants grown in replicated, comparative trials under uniform conditions. (If the entire experiment is repeated in several different sets of conditions estimates can be made of the interaction between the population genotype and the planting environment.) These should be supported by studies of material from the natural range also. Foresters have established many thousands of reliable, replicated provenance trials of the major plantation species in tropical, temperate and arid/semi-arid regions but relatively few of these have been studied anatomically other than for the major features known to be correlated with timber use characteristics. Examples of the analysis and use of such material are discussed below but the attention of anatomists is drawn here to the availability of these experiments managed by governments, universities, commercial companies and research institutions in many countries.

Collection of seeds and assessment of field experiments is often coordinated by the International Union of Forestry Research Organizations and several international comparative trials, mainly for the benefit of developing and tropical countries, have been established with the assistance of Commonwealth Forestry Institute, Oxford, England (Central American tropical pines, *Cedrela, Cordia*), Commonwealth Scientific and Industrial Organization, Canberra, Australia (*Eucalyptus* species), Danish-FAO Tree Seed Centre, Humlebaek, Denmark (*Gmelina, Tectona,* Asian tropical pines). Centre Technique Forestier Tropical, Paris, France, and Instituto Nacional de Investigaciones Forestales, Mexico, have also collaborated in some of the collections and FAO is currently organising international collection and testing of *Acacia* and *Prosopis* species for arid and semi-arid zones. In temperate countries international research has concentrated on North West American and European plantation conifers including particularly *Abies grandis, Larix europaea, Picea abies, P. sitchensis, Pinus contorta, P. sylvestris* and *Pseudotsuga menziesii.*

Choice of characters

Programmes of tree breeding developed significantly from 1950 onwards and today many countries have programmes of selection, testing and propagation of populations and individual genotypes. Although the principal traits selected relate to vigour, pest resistance, stem form and crown characters, several wood properties are included, either in the initial phenotypic selection or in subsequent genetic evaluation (progeny testing). Because of the cost of assessing the large numbers of samples required for adequate precision in any wood property and because of the difficulty of evaluating some technological properties (*e.g.* timber strength or pulp and paper characters), considerable attention has been paid to correlations between these traits and more simply assessed anatomical features.

   Several anatomical characters or their derivatives are related to a greater or lesser extent to technological properties and several relevant reviews and bibliographies of variation and correlation have been prepared. These may not be familiar to wood anatomists and some are listed here for their convenience, including:

   wood density in conifers (Elliott, 1970; Spurr and Hsiung, 1954);
   variation of tracheid and fibre length (Spurr and Hyvarinen, 1954; Dinwoodie, 1961);
   spiral grain (Noshowiak, 1963);
   juvenile wood (Paul, 1957; Polge, 1964; Zobel, 1980);
   compression wood (Low, 1964; Westing, 1965, 1968);
   tension wood (Hughes, 1965);

relations between fibre morphology and paper properties (van Buijtenen, 1969; Dinwoodie, 1965);

influence of silvicultural practices on wood properties (Fielding, 1967);

effects of fertilisation on wood quality (Cleaveland and Wooten, 1974);

genotype-environment interaction effects on wood properties of conifers (Goggans, 1961);

genetic and environmental effects on pulpwood quality (Kouris, 1962; Weiner and Roth, 1966);

growth stress (Bamber, 1978b; Dinwoodie, 1966).

If wood quality is considered as the suitability of wood for use, the best single indicator of quality is wood density; this feature is a man-made concept dependent on several anatomical features, yet it has been shown to be heritable in many species and reasonably closely related to properties of timber sawing, machining, gluing, preservation, strength and pulping in most. Among the anatomical features that contribute to density, earlywood/latewood proportions have been studied widely, cell wall thickness/lumen diameter ratios have been studied for the major pulpwood species, and fibre/vessel/parenchyma proportions have been examined for some species in relation to pulp manufacture and for many others in connection with taxonomic and descriptive research. However, for all these traits there is a dearth of information on the extent and pattern of genetic variation.

As geneticists and silviculturists increase the growth rates of plantations and as the demand for timber increases, the age at which plantations are felled for use is declining with a consequent increase in the proportion of juvenile wood being marketed. Since the within-tree variation of density (and fibre length, cell wall thickness and grain angle) is greatest in the juvenile core there is an increasing opportunity for geneticists to improve wood quality in this region (see *e.g.* Zobel, 1976; Zobel *et al.,* 1978).

Apart from any contributions to density, fibre dimensions *per se* influence timber strength and pulp processes and properties. Tracheid length in conifers is usually acceptable for Kraft pulping but fibres in hardwoods are too short (less than 2 mm) and would require a probably unapproachable genetic improvement of 50 per cent in this basic property to bring them into this higher value catagory. (However, bags, newsprint and other papers are made now from short-fibred material including eucalypts.) Cell wall thickness (and the ratio of cell wall thickness to lumen or total cell diameter) also contribute to pulp and paper characteristics (burst, break, tear and folding) and hence the proportion of compression wood is of interest.

For sawn timber, grain inclination ('spiral grain') is of importance and commonly included in tree breeding programmes for conifers. The microfibrillar

156

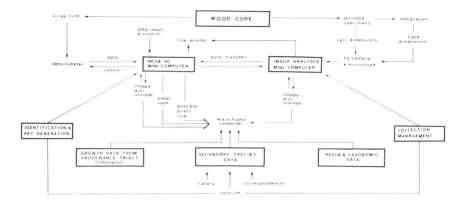

Fig. 1. Anatomical data gathering, processing, storage and retrieval at Commonwealth Forestry Institute, Oxford, England.

(micellar) angle within cell walls is known to influence strength but is more difficult to evaluate.

In summary it can be said that, whereas density and its components, fibre dimensions and orientation have been studied genetically because they influence wood quality, the genetic variations of many other anatomical features of interest to wood anatomists have been largely ignored.

Sampling and evaluation

Because of the large numbers of samples required to estimate mean values precisely, considerable research is needed for any new species to determine optimum sampling strategy (see Burley and Miller, 1982). In the past 50 years the trends have been from large, industrial scale testing of processing methods and technological properties for large timbers or pulp samples selected at random from commercial supplies to laboratory scale examination of anatomical features in small wood samples (*e.g.* radial increment cores 5, 8, 11 or 12 mm in diameter) with a sampling system designed to represent the variation adequately (see *e.g.* Forest Biology Subcommittee No. 2, 1968; Mitchell, 1958; Van der Sijde, 1979).

For anatomical assessment, trends have been away from the laborious human measurement of gravimetric density and projected images of single fibres to automated photometric, densitometric and image-analysis methods (Fletcher and Hughes, 1970; Harris and Polge, 1967; Hughes and Sardinha, 1975; Phillips, 1966; Polge *et al.*, 1977; Polge and Nicholls, 1972).

As computational methods and electronic technology improve, more interfacing of equipment is possible and more automatic evaluation of samples, analysis of data and storage/retrieval of information is developing. It is a large jump from the single microscope and technician that supported Dr. L. Chalk in his massive contribution to the 'Anatomy of the Dicotyledons' (Metcalfe and Chalk, 1950) to the four professional and four technical staff that operate the network of high speed densitometer, image analyser and two computers now installed at Oxford (see Fig. 1). There is always a danger that modern technology may be used to replace classical intellectual ability but there is no doubt that modern methods are necessary to process the large numbers of samples and pieces of information that are needed to study genetic variation in many wood properties and species.

Geographic variation between populations

It is not possible to review all the studies of wood properties in natural populations and provenance trials that have been published. The majority of the early reports (the 1960s decade) dealt with wood density, with pines and with North America and showed latitudinal and altitudinal trends among natural populations, *e.g.* southern pines (Mitchell, 1964; Forest Products Laboratory, 1965a; Wahlgren *et al.*, 1975), western species (Forest Products Laboratory, 1965b), longleaf pine in Virginia (Saucier and Taras, 1966), white pine and red pine in Illinois (Gilmore, 1968), slash pine (Goddard and Strickland, 1962), Virginia pine (Thor, 1964) and loblolly pine (Zobel *et al.*, 1960). Many more examples could be quoted but these are sufficient to show that geographic, habitat-correlated variation occurs. Several of these studies but particularly the initial study by Zobel *et al.* (1960) showed that there was considerable additional variation attributable to forest stands (sites) within geographic locations and to individual trees within populations. Several studies were of provenance trials that demonstrated genetic control to varying degrees and one showed genetic variation in densitometric properties (but not in tracheid length or spiral grain) between three mainland and two insular provenances of radiata pine grown in five Australian provenance trials (Nicholls and Eldridge, 1980).

Examples of other coniferous characteristics and species studied include tracheid length in Scots pine (Echols, 1958, in provenance trials in eastern U.S.A.; Miler *et al.*, 1979, in Poland), shortleaf pine (Posey *et al.*, 1970, in Oklahoma, U.S.A.), black pine (Lee, 1979, in Michigan, U.S.A.); jack pine (Kennedy, 1971, in Canada), tangential tracheid diameter in western red cedar (Crooks, 1966) and tracheid length, insoluble carbohydrates, *a*-cellulose and compression wood in loblolly pine (Zobel *et al.*, 1960, in seven States of the U.S.A.).

Evidence of interaction between population genetic effects and the environment of the provenance trial location was given by Echols (1973) and Kennedy (1971) among others and there have been many reports of such effects on seedling anatomy in trials in nursery, greenhouse and growth chamber. Occasionally no significant variation is detected among natural populations, *e.g.* Strickland and Goddard (1966) found no differences in tracheid length between 15 geographic areas of slash pine even though great differences occurred for wood density (Goddard and Strickland, 1962). No variation between populations appeared to exist for specific gravity of red alder in Canadian provenance trials (Harrington and DeBell, 1980).

Apart from the studies related to ecological anatomy discussed above there have been fewer reports of geographic variation in dicotyledonous species. Examples include specific gravity and fibre length of eastern cotton wood in the southern Great Plains of the U.S.A. (Posey *et al.*, 1969), various species in the mid-south of the U.S.A. (Sluder, 1972; Taylor, 1977), northern red oak in the U.S.A. (Maeglin, 1976), white ash in a provenance trial in Illinois, U.S.A. (Armstrong and Funk, 1980) and sweetgum in a controlled environment (Winstead, 1967). Geographic variation in wood quality of *Tectona grandis*, particularly density, has been studied in natural stands and provenance trials (Kedharnath *et al.*, 1963; Purkayastha and Satyamurthi, 1975; Purkayastha *et al.*, 1973). Webb (1965) reported variation in interlocked grain among 225 trees of *Liquidambar styraciflua* in the southern Atlantic coast States of the U.S.A. and Maeglin (1976) also recorded natural variation in tissue proportions of northern red oak. In seven-year thinnings of *Eucalyptus obliqua* in Australia, significant differences were observed between 14 provenances in heartwood proportion, incidence of kino veins, latewood ratio and density variation with rings (Nicholls and Matheson, 1980).

Occasionally a tree species is introduced for plantation without preliminary provenance testing to determine the optimum seed source and with no indication of total variability. While awaiting the establishment and assessment of provenance trials one stage is to determine the variability of plantations of one seed lot among a range of sites and management conditions. This was the approach of Akachuku and Burley (1979) with *Gmelina arborea,* a promising Indian pulpwood species in Nigeria; they found that four principal sites varied only for fibre length but that individual trees within sites varied significantly for fibre length, and for proportions of fibres, vessels and parenchyma. It was also the method used in a long series of studies at the Commonwealth Forestry Institute, Oxford, England, of tropical pines (particularly *Pinus caribaea* and

P. oocarpa) in their native habitat and in plantations at several locations throughout the tropics (including Fiji, Jamaica, Kenya, Tanzania, Trinidad, and Uganda).

Variation between individual trees

In most of the studies of geographic variation, whether based on natural forests or provenance trials, a common conclusion is that individual trees vary within a given population. This is true equally for anatomical and morphological properties and most features appear to be under polygenic control with varying degrees of environmental effects. If a normal distribution is assumed, genetic advance in a given feature depends on selection intensity and heritability; the tree breeder needs to know the extent of variation in the population within which selection is to be made and the proportion of that variation that is under genetic control.

The ratio of genetic variance to total phenotypic variance is termed the heritability and it is strictly relevant only to the particular population, environment and characteristic for which it is determined. It can be expressed in two forms; broad sense heritability is the ratio of total genetic variance to total phenotypic variance while narrow sense heritability is the ratio of additive genetic variance to total phenotypic variance. The former is used when trees are propagated vegetatively (clonally) without sexual reproduction, meiosis and genetic recombination. The latter refers to seedling propagation in which only additive genetic effects are useful; dominance and epistatic effects can only be used with special breeding procedures.

Heritability is always smaller in the narrow sense than in the broad sense and thus greater genetic gains are possible with clonal selection and propagation for a given selection intensity. In addition to several industrial woody crops such as rubber, tea, cocoa, many fruit trees and horticultural species, several forest tree species are propagated vegetatively including some tropical eucalypts, teak, and some conifers including *Picea abies, Pseudotsuga menziesii,* and *Pinus radiata.*

While genetic gain depends also on the variation present and the selection intensity applied, heritability gives an indication of the genetic advance possible and a large number of heritability estimates for various anatomical and chemical characters have been published (see summaries in Einspahr, 1972; Hattemer, 1963; Nicholls, 1978; Smith, 1967; Zobel 1961, 1964, 1971). These have been based on analysis of variance of seedlings derived from one-parent or two-parent controlled crosses, from parent-offspring regression and from clonal tests; the numbers of families and replications differ and thus the precision of heritability estimates varies. However, the size of the heritability has led to the inclusion of several traits in breeding programmes with proven genetic gain for some, e.g. *Pinus elliottii* in Queensland (Dadswell and Nicholls, 1959) *Pinus pinaster* in Western Australia (Nicholls, 1967a); *Pinus patula* and *Cupressus lusitanica* in East Africa (Paterson, 1968). The most intensive work has been with the southern pines in the North Carolina State University – Industry Cooperative Program, U.S.A., led by Zobel and described in the Program's annual reports and several of Zobel's papers cited here (see Namkoong et al., 1969; Zobel, 1970).

These have been mainly with conifers, especially southern pines in the U.S.A. and particularly have dealt with wood density (narrow sense heritability, h^2n.s. = 0.2—0.7) and tracheid length (h^2n.s. = 0.3—0.9) but other characteristics include cell diameter and wall thickness in *Pinus taeda* (h^2n.s. = 0.13—0.84, Goggans, 1964), cellulose content for *P. radiata* (Dadswell et al., 1961) and *P. taeda* (largely dominance genetic variance, Jett et al., 1977; Zobel et al., 1966), tracheid length for *P. elliottii* (Echols, 1955), proportion of compression wood in *P. taeda* (Shelbourne et al., 1969), spiral grain (Dadswell et al., 1961; Zobel, 1961, 1965a; Zobel et al., 1968) and proportion of heartwood for *P. radiata* (Nicholls and Brown, 1974).

For dicotyledonous species there is less information on genetic control within populations other than for some plantation eucalypts grown in Australia, Brazil, South Africa and the southern U.S.A. The few early references to inheritance of grain and figure in various species and fibre length in poplars were reviewed by Zobel (1964, 1965a, b). More recently in three year *Eucalyptus viminalis* grown in Georgia, U.S.A., family heritabilities were found to be 0.4—0.9 for wood density, moisture content, bark density, fibre length, cell wall thickness, fibre diameter and lumen diameter (Otegbeye and Kellison, 1980). Spiral grain in *Fagus sylvatica* (h^2n.s. = 0.66) showed strong additive genetic control (Teissier du Cros et al., 1980).

Nine American clones of *Populus*, grown under short rotation intensive culture, differed significantly in wood and bark density, vessel member length and fibre length indicating the range of variation available to the breeder for this type of silvicultural system (Crist, 1980). Large differences were found between 18 American hybrid clones of *Populus* in content of lignin, glucose, xylose, mannose and arabinose (Dickson et al., 1974); some clones combined fast growth with low lignin and high glucose content which would be desirable commercially.

Variation and correlation within trees

In the preceding discussions population and tree means were assumed to be adequately assessed but this ignores variation within trees. Since the original enunciation of Sanio's laws (Sanio, 1872) it has been demonstrated for a wide range of species that fibre length, density, grain angle and other features have characteristic patterns of variation along the radius and along the stem. The pattern varies with character, species and tree but there have been few attempts to derive adequate sampling strategies (see Burley *et al.*, 1970) or quantitative models (*e.g.* Burley and Andrew, 1970; Kandeel, 1971; Taylor, 1968). It is desirable to know the parameters of these models before valid comparisons can be made between trees, sites or management treatments.

While the heritability of a wood property may vary with the age of the tree or the annual ring in which the wood is formed (*e.g. Pinus radiata* density, Nicholls, 1976b), the variation within trees or even within annual rings may be of value. Thus Zobel *et al.* (1978) obtained genetic gains of 32 kg m⁻³ in the density of juvenile wood of southern pines (see also Zobel, 1970) and Burley and Palmer (1979) found that intra-ring densitometric variability of *Pinus caribaea* added to the efficiency of mean tree density in predicting pulp properties.

Knowledge of within-tree variation patterns is required for prediction of mature properties from juvenile samples. Most commercial plantations are raised on long rotations of 15-60 years, and, in conservation and breeding programmes, there is urgent need for early evaluation (Burley, 1981; Lambeth, 1980). Unfortunately in wood properties these correlations are sometimes poor; for example, in a 46 year old seed source trial of *Pseudotsuga menziesii* in Washington State, U.S.A., juvenile and core wood were unreliable for predicting the density of mature wood which was in fact better correlated with date of phenological flushing (McKimmy, 1966; McKimmy and Nicholas, 1971).

The precise determination of correlations needs adequate sampling and this was lacking in the studies of three clones of *Populus* by Sacré (1977); nevertheless the quality of adult wood was indicated by several juvenile traits including proportion of tension wood, number of vessels per square millimetre, fibre diameter, density and shrinkage on drying.

In *Liquidambar styraciflua* the distribution patterns of vessels, fibres and parenchyma were studied within trees (Ezell, 1979). Regression analyses using fibre or vessel proportions predicted wood density precisely. (It is also of interest that fibre proportion increased and vessel proportion decreased with increasing height while parenchyma proportion remained unchanged but this may be due to the major problem in studies of within-tree variation, namely that wood at the top of trees is physically and physiologically younger than material at the bottom.)

Such correlations between traits within trees (after adjustment for age or size effects) may have adverse effects in selection since the improvement of one character may result in the deterioration of another. Taylor (1969), for example, suggested that, since ray tissue is denser than surrounding tissue, selection for wood density may increase the proportion of ray tissue which would be undesirable for many end uses.

Hybrids and polyploids

Only brief mention is made here of two types of tree that may occur naturally and that are of importance in some tree improvement programmes. Hybrids have commercial importance particularly in eucalypts, larches, pines, poplars and willows. Hybrid vigour or superiority is sought in wood properties as in growth rates, tree form and disease or climatic resistance; it may occur for some anatomical features but not others in the same hybrid (*e.g.* Koo and Hong, 1966, reported that hybrids and backcrosses between *Pinus rigida* and *P. taeda* in Korea had greater tracheid length but smaller tracheid width and compression strength than the parents). In 28 year old hybrids between *Larix decidua* and *L. leptolepis* hybrid vigour was apparent only for the first ten years in ring width, wood density, extractive content and the ratio of latewood and earlywood densities (Reck, 1980).

Although polyploids have been induced artificially particularly with colchicine in many species (*e.g.* Otsuka *et al.*, 1964, reported that 21 year old tetraploid *Pinus thunbergii* in Japan had shorter tracheids with thicker walls than the diploid) the main genus of commercial interest is *Populus, e.g.* in the U.S.A. the hybrid aspen, *Populus tremuloides* in which significant differences in growth rate, wood density and fibre length occur (Einspahr *et al.*, 1968). Recent work in Russia on several poplar species and hybrids has shown that several anatomical features, particularly ray characteristics, are suitable for identifying polyploids (Bakulin, 1980); compared to diploids, tetraploids had larger cells, fewer vessels per unit area, and rays with a reduced density, height, number of cell rows and number of horizontal rows of pits on their side walls.

Current needs and future developments

As wider exploration and more intensive exploitation of remaining natural forests proceed, and as more species are introduced for commercial plantations, there is urgency in assessing the extent and pattern of variation in anatomical features. Ecologists, taxonomists and physiologists can benefit directly from such information and forest managers and wood users can benefit indirectly. Taxonomists and tree breeders need to determine the contribution of genetic variation between populations and between trees to the total phenotypic variation observed within a species and to the correlations detected between anatomical and utilisation characteristics.

Straightforward description of the anatomy of wood from natural forest or commercial plantations, with or without quantitative models of intra-tree variation, forms a basic source of information. However, more experimental work is needed to relate the variation to environmental variables in the field (see *e.g.* Denne, 1979, for a study of tracheids of *Picea sitchensis* in relation to the distribution of light intensity and leaf weight as the tree canopy developed) or in controlled environments such as growth chambers, greenhouses or forest nurseries (see *e.g.* Rudman, 1970, for a study of fibre length in clonal material of *Eucalyptus camaldulensis* grown in two phytotron glasshouses).

While the wood anatomist may continue to be more interested in describing anatomical features *per se*, future tree breeders may be interested in detecting, describing and changing characters and processes that give rise to desirable anatomical properties. Genetic engineering in forest trees is likely to concentrate initially on introduction of nitrogen fixation into non-legumes and on root mycorrhizae, but improvement of growth by indirect selection for rates of physiological processes such as photosynthesis or respiration is already being considered (see papers in Cannell and Last, 1976); it is thus conceivable that selection for molecular processes influencing cell wall growth and structure (see *e.g.* Chafe, 1978; Berlyn, 1979) may be feasible in the future.

Conclusions

The main conclusions to be drawn from the extensive literature on genetic variation and breeding for wood properties are:
1) genetic gain is possible for a range of characters through selection between and within populations,
2) information on genetic variance and heritability is always specific to population, environment, management and character,
3) information is restricted largely to wood density, fibre dimensions and grain inclination, particularly in conifers with relatively little for dicotyledons, and
4) little use is made by wood anatomists of the large number of comparative provenance and progeny trials that facilitate estimation of the genetic component of variation within and between populations.

References

Akachuku, A.E. & J. Burley. 1979. Variation of wood anatomy of Gmelina arborea Roxb. in Nigerian plantations. IAWA Bull. 1979/4: 94-99.

Armstrong, J.E. & D.T. Funk. 1980. Genetic variation in the wood of Fraxinus americana. Wood and Fiber 12: 112-120.

Baas, P. 1973. The wood anatomical range in Ilex (Aquifoliaceae) and its ecological and phylogenetic significance. Blumea 21: 193-258.

Baas, P. 1976. Some functional and adaptive aspects of vessel member morphology. In: P. Baas, A.J. Bolton & D.M. Catling (eds.), Wood Structure in Biological and Technological Research: 157-181. Leiden Bot. Series No. 3 Leiden Univ. Press, The Hague.

Bakulin, V.T. 1980. Wood anatomy of induced polyploids of poplar. Izv. Sibirsk. Otdel. Akad. Nauk SSSR, Biol. Nauk No. 1 (5): 25-30.

Bamber, R.K. 1978a. Wood anatomy - a fundamental branch of wood science. Pap. IUFRO Conf. Wood Quality and Utilization of Tropical Species, Laguna, Philippines, 1978: 13-28.

Bamber, R.K. 1978b. The origin of growth stresses. Pap. IUFRO Conf. Wood Quality and Utilization of Tropical Species, Laguna, Philippines, 1978: 7 pp.

Berlyn, G.P. 1979. Physiological control of differentiation of xylem elements. Wood and Fiber 11: 109-126.

Bormann, F.H. & G.P. Berlyn (eds.). 1981. Age and growth rate of tropical trees. Yale Univ. School For. Environm. Stud. Bull. No. 94: 137 pp.

Brazier, J.D. 1968. The contribution of wood anatomy to taxonomy. Proc. Linn. Soc. Lond. 179: 271-274.

Buijtenen, J.P. van. 1969. Relationships between wood properties and pulp and paper properties. Pap. 2nd World Consultn. For. Tree Breed., Washington, D.C., U.S.A., FO-FBT-69-4/5, FAO, Rome, Italy.

Burley, J. 1981. Time-related problems in the evaluation of forest genetic resources. Pap. FAO/UNEP/IBPGR Techn. Conf. Crop Genetic Resources, Rome, Italy: 14 pp.

Burley, J., P.G. Adlard & P. Waters. 1970. Variances of tracheid lengths in tropical pines from central Africa. Wood Sci. Technol. 4: 36-44.

Burley, J. & I.A. Andrew. 1970. Variation in wood properties of Pinus kesiya Royle ex Gordon (syn. P. khasya Royle; P. insularis Endlicher); six trees of Assam provenance grown in Zambia. Wood Sci. Technol. 4: 195-212.

Burley, J. & R.B. Miller. 1982. The application of statistics and computing in wood anatomy. This volume: 223-242.

Burley, J. & E.R. Palmer (principal investigators). 1979. Pulp and wood densitometric properties of Pinus caribaea from Fiji. CFI Occ. Paper No. 6: 66 pp. Oxford, England.

Cannell, M.G.R. & F.T. Last. 1976. Tree physiology and yield improvement. Acad. Press, London.

Carlquist, S. 1975. Ecological strategies of xylem evolution. Univ. California Press, Berkeley.

Chafe, S.C. 1978. On the mechanisms of cell wall microfibrillar orientation. Wood Sci. Technol. 12: 203-217.

Champion, H.G. 1925. Contributions towards a knowledge of twisted fibre in trees. Indian For. Rec. (Silvic. Series) 11 (2): 11-80.

Champion, H.G. 1933. Importance of the origin of seed used in forestry. Indian For. Rec. (Silvic. Series) 17 (5): 11, 14-18.

Cleaveland, M.K. & T.E. Wooten. 1974. Annotated bibliography on the effects of fertilization on wood quality. For. Res. Series, Clemson Univ., South Carolina, No. 29: 53 pp.

Crist, J.B. 1980. Anatomical characteristics of nine Populus clones grown under short rotation intensive culture. Abstr. Pap. 34th Ann. Mtg. For. Prod. Res. Soc., Boston, U.S.A.: 6-7.

Crooks, J.R. 1966. Geographic variation of tangential tracheid diameter in Idaho western red cedar. Abstr. of thesis in For. Prod. J. 16 (11): 55.

Dadswell, H.E., J.M. Fielding, J.W.P. Nicholls & A.G. Brown. 1961. Tree-to-tree variation and the gross heritability of wood characteristics of Pinus radiata. TAPPI 44: 174-179.

Dadswell, H.E. & J.W.P. Nicholls. 1959. Assessment of wood qualities for tree breeding. 1. In Pinus elliottii var. elliottii from Queensland. Div. For. Prod. CSIRO Melbourne, Australia, Techn. Pap. 4: 66 pp.

Denne, M.P. 1979. Wood structure and production within the trunk and branches of Picea sitchensis in relation to canopy formation. Canad. J. For. Res. 9: 406-427.

Dickison, W.C., R.M. Rury & G.L. Stebbins. 1978. Xylem anatomy of Hibbertia (Dilleniaceae) in relation to ecology and evolution. J. Arn. Arbor. 59: 32-49.

Dickson, R.E., P.R. Larson & J.G. Isebrands. 1974. Differences in cell-wall chemical composition among eighteen three-year-old Populus hybrid clones. In: Proc. 9th Central States For. Tree Impr. Conf., Ames, Iowa, U.S.A.: 21-34.

Dinwoodie, J.M. 1961. Tracheid and fibre length in timber – a review of literature. Forestry 34: 125-144.

Dinwoodie, J.M. 1965. The relationship between fiber morphology and paper properties – a review of literature. TAPPI 48: 440-447.

Dinwoodie, J.M. 1966. Growth stress in timber – a review of literature. Forestry 39: 162-170.

Echols, R.M. 1955. Linear relation of fibrillar angle to tracheid length, and genetic control of tracheid length in slash pine. Trop. Woods 102: 11-22.

Echols, R.M. 1958. Variation in tracheid length and wood density in geographic races of Scotch Pine. Yale Univ. School For. Bull. No. 64: 52 pp.

Echols, R.M. 1973. Effects of elevation and seed source on tracheid length in young ponderosa pine. For. Sci. 19: 46-49.

Einspahr, D.W. 1972. Forest tree improvement and the pulp and paper industry. TAPPI 55: 1660-1669.

Einspahr, D.W., M.K. Benson & J.R. Peckham. 1968. Wood and pulp properties of 5-year-old diploid and triploid hybrid aspen. TAPPI 51: 72-75.

Elliott, G.K. 1970. Wood density in conifers. Techn. Commun. Commonw. For. Bur. No. 8: 44 pp.

Ezell, A.W. 1979. Variation of cellular proportions in sweetgums and their relation to other wood properties. Wood and Fiber 11: 136-143.

Fahn, A., J. Burley, K.A. Longman, A. Mariaux & P.B. Tomlinson. 1981. Possible contributions of wood anatomy to the determination of age of tropical trees. In: F.H. Bormann & G.P. Berlyn (eds.), Age and Growth Rate of Tropical Trees: 31-54. Yale Univ. School For. Environm. Stud. No. 94.

Fielding, J.M. 1967. The influence of silvicultural practices on wood properties. Int. Rev. For. Res. 2: 95-126.

Fletcher, J.M. & J.F. Hughes. 1970. Uses of X-rays for density determinations and dendrochronology. Bull. Fac. For. U.B.C. 7: 41-54.

Forest Biology Subcommittee No. 2. 1968. New methods of measuring wood and fiber properties in small samples. TAPPI 51: 75A-80A.

Forest Products Laboratory. 1965a. Southern wood density survey – 1965 status report. U.S. For. Serv. Pap. F.P.L. 26: 38 pp.

Forest Products Laboratory. 1965b. Western wood density survey report No. 1. U.S. For. Serv. Res. Pap. F.P.L. 27: 58 pp.

Gilmore, A.R. 1968. Geographic variations in specific gravity of white pine and red pine in Illinois. For. Prod. J. 18 (11): 49-51.

Goddard, R.E. & R.K. Strickland. 1962. Geographic variation in wood specific gravity of slash pine. TAPPI 45: 606-608.

Goggans, J.F. 1961. The interplay of environment and heredity as factors controlling wood properties in conifers. N.C. State Coll. Techn. Rep. 11: 56 pp.

Goggans, J.F. 1964. Correlation and heritability of certain wood properties in loblolly pine (Pinus taeda L.). TAPPI 47: 318-322.

Graaff, N.A. van der, & P. Baas. 1974. Wood anatomical variation in relation to latitude and altitude. Blumea 22: 101-121.

Hamilton, T. 1761. Treatise on the manner of raising forest trees, in a letter from the Right Honourable, the Earl of Haddington to his grandson. To which are added two memoirs; the one on preserving and repairing forests; the other on the culture of forests. Both translated from the French of M. Buffon. Printed for G. Hamilton, Edinburgh: 129 pp.

Harrington, C.A. & D.S. DeBell. 1980. Variation in specific gravity of red alder (Alnus rubra Bong.). Canad. J. For. Res. 10: 293-299.

Harris, J.M. & H. Polge. 1967. A comparison of X-ray and beta-ray techniques for measuring wood density. J. Inst. Wood Sci. 19: 34-42.

Hattemer, H.H. 1963. Estimates of heritability published in forest tree breeding research. Pap. FAO World Consultn. For. Genet., Stockholm, FAO/FORGEN 63/2a/3: 14 pp.

Heywood, V.H. (ed.). 1973. Taxonomy and ecology. Proceedings of an international symposium held at the Department of Botany, University of Reading. Spec. Vol. Syst. Assoc. Acad. Press, London, No. 5: 370 pp.

Hughes, J.F. 1965. Tension wood: a review of literature. Part I. Occurrence and development of tension wood. Part II. The properties and use characteristics of tension wood. For. Abstr. 26: 2-9; 179-186.

Hughes, J.F. & R.M. de A. Sardinha. 1975. The application of optical densitometry in the study of wood structures and properties. J. Microscopy 104: 91-103.

Jett, J.B., R.J. Weir & J.A. Barker. 1977. The inheritance of cellulose in loblolly pine. In: Proc. TAPPI For. Biol. Conf.: 159-162.

Kandeel, E.S.A. 1971. Polynomial models to study and present within-tree variation of wood properties. Wood and Fiber 3: 106-111.

Kedharnath, S., V.J. Chacko, S.K. Gupta & J.D. Matthews. 1963. Geographic and individual tree variation in some wood characters of teak (Tectona grandis L. f.) Silvae Genetica 12: 181-186.

Kennedy, R.W. 1971. Influence of provenance and test location on wood formation in Pinus banksiana Lamb. seedlings. Silvae Genetica 20: 60-63.

Koehler, A. 1939. Heredity versus environment in improving wood in forest trees. J. For. 37: 683-687.

Koo, K.H. & S.H. Hong. 1966. Tracheid length and compression strength of x Pinus rigitaeda and its back-cross hybrid. Res. Reports, Office of Rural Devel., Suwon 9: 117-123.

Kouris, M. (ed.). 1962. The influence of environment and genetics on pulpwood quality. TAPPI Monograph 24: 316 pp.

Lambeth, C.C. 1980. Juvenile-mature correlations in Pinaceae and implications for early selection. For. Sci. 26: 571-580.

Lee, C.H. 1979. Absence of growth-wood property correlation in twenty-seven black pine seed sources. Wood and Fiber 11: 22-28.

Low, A.J. 1964. Compression wood in conifers. A literature review. Parts I and II. For. Abstr. 25 (3): xxxv-xliii and 25 (4): xlv-li.

Maeglin, R.R. 1976. Natural variation of tissue proportions and vessel and fiber length in mature red oak. Silvae Genetica 25: 122-126.

McKimmy, M.D. 1966. A variation and heritability study of wood specific gravity in 46-year old Douglas Fir from known seed sources. TAPPI 49: 542-549.

McKimmy, M.D. & D.D. Nicholas. 1971. Genetic differences in wood traits among half-century-old families of Douglas Fir. Wood and Fiber 2: 347-355.

Metcalfe, C.R. & L. Chalk. 1950. Anatomy of the Dicotyledons. Vol. 1 & 2. Clarendon Press, Oxford.

Miler, Z., A. Miler & P. Pasternak. 1979. Wood fibre length in pine provenance samples. Prace Kom. Nauk Roln. i Kom. Nauk Lesnych No. 48: 95-101.

Miller, R.B. 1976. Wood anatomy of the species of Juglans. Bot. Gaz. 137: 368-377.

Mitchell, H.L. 1958. Wood quality evaluation from increment cores. TAPPI 41: 150-156.

Mitchell, H.L. 1964. Patterns of variation in specific gravity of southern pines and other coniferous species. TAPPI 47: 216-283.

Namkoong, G., A.C. Barefoot & R.G. Hitchings. 1969. Evaluating control of wood quality through breeding. TAPPI 52: 1935-1938.

Nicholls, J.W.P. 1967a. Assessment of wood qualities for tree breeding. IV. Pinus pinaster Ait. grown in Western Australia. Silvae Genetica 16: 21-28.

Nicholls, J.W.P. 1967b. Preliminary observations on the change with age of the heritability of certain wood characters in Pinus radiata clones. Silvae Genetica 16: 18-20.

Nicholls, J.W.P. 1978. A review of the genetic parameters of some wood characteristics of Pinus radiata. Pap. 6th Meet. Res. Work Group 1, Aust. For. Counc., Coff's Harbour, Australia: 14 pp.

Nicholls, J.W.P. & A.G. Brown. 1974. The inheritance of heartwood formation in Pinus radiata D. Don. Silvae Genetica 23: 138-141.

Nicholls, J.W.P. & K.G. Eldridge. 1980. Variation in some wood and bark characteristics in provenances of Pinus radiata D. Don. Austr. For. Res. 10: 321-335.

Nicholls, J.W.P. & A.C. Matheson. 1980. Variation in wood characteristics in thinnings from a field trial of Eucalyptus obliqua. Austr. For. Res. 10: 239-247.

Noshowiak, A.F. 1963. Spiral grain in trees – a review. For. Prod. J. 13: 266-275.

Novruzova, Z.A. 1968. The water-conducting system of trees and shrubs in relation to ecology. Izv. Akad. Nauk. Azerb. SSR, Baku: 230 pp.

Oever, L. van den, P. Baas & M. Zandee. 1980. Quantitative wood anatomy and provenance in the genus Symplocos. Pap. Proc. IUFRO Wkg. Pty. S5.01-02, Oxford, U.K.; Mitt. Bundesforschungsanst. f. Forst- u. Holzwirtschaft, Hamburg, Germany, No. 131: 49-60.

Oever, L. van den, P. Baas & M. Zandee. 1981. Comparative wood anatomy of Symplocos and latitude and altitude of provenance. IAWA Bull. n.s. 2: 3-24.

Otegbeye, G.O. & R.C. Kellison. 1980. Genetics of wood and bark characteristics of Eucalyptus viminalis. Silvae Genetica 29: 27-31.

Otsuka, M., S. Toyama & M. Yonemochi. 1964. On the characteristics of tetraploid Pinus thunbergii. Bull. Fac. Agric. Univ. Miyazaki 10: 153-161.

Outer, R.W. den, & W.L.H. van Veenendaal. 1976. Variation in wood anatomy of species with a distribution covering both rain forest and savanna areas of the Ivory Coast, West-Africa. In: P. Baas, A.J. Bolton & D.M. Catling (eds.), Wood Structure in Biological and Technological Research: 182-195. Leiden Bot. Series No. 3. Leiden Univ. Press, The Hague.

Paterson, D.N. 1968. Control of wood quality and quantity in the east African exotic softwood tree breeding programme. E. Afr. Ag. & For. J. 33: 302-315.

Paul, B.H. 1957. Juvenile wood in conifers. U.S. Forest Serv., For. Prod. Lab. Report 2094: 6 pp.

Phillips, E.W.J. 1966. The use of beta-particle radiation methods in timber research. News Bull. IAWA 1966/2: 17-28.

Polge, H. 1964. Le bois juvenile des conifères. Rev. For. Franc. 16: 474-505.

Polge, H., M. Lemoine & E. Deret. 1977. A study of specific and infraspecific variability in the structure of juvenile wood of oak by means of an image analyser. Ann. Sci. For. 34: 285-292.

Polge, H. & J.W.P. Nicholls. 1972. Quantitative radiography and the densitometric analysis of wood. Wood Science 5: 51-59.

Posey, C.E., F.E. Bridgwater & J.A. Buxton. 1969. Natural variation in specific gravity, fiber length, and growth rate of eastern cottonwood in the southern Great Plains. TAPPI 52: 1508-1511.

Posey, C.E., F.E. Bridgwater & N. Walker. 1970. Effect of seed origin on tracheid length, specific gravity, and volume of shortleaf pine in Oklahoma. For. Sci. 16: 66-70.

Purkayastha, S.K. & K.R. Satyamurthi. 1975. Relative importance of locality and seed origin in determining wood quality in teak. Indian For. 101: 606-607.

Purkayastha, S.K., R.D. Tandon & K.R. Rao. 1973. A note on the variation in wood density in some 36-year-old teak trees from different seed origins. Indian For. 99: 215-217.

Reck, S. 1980. Analysis of the wood of hybrid larches. Allg. Forst- u. Jagdzeitung 151: 117-120.

Rudman, P. 1970. The influence of genotype and environment on wood properties of juvenile Eucalyptus camaldulensis Dehn. Silvae Genetica 19: 49-54.

Sacré, E. 1977. Caractéristiques anatomiques et physiques du bois des peupliers 'I. 214', 'robusta' et 'gelrica' aux stades prócoce et adulte. Bull. Soc. Roy. For. Belgique 84: 321-338 (French, with Dutch summary, 4 refs.).

Sanio, K. 1872. Über die Grösse der Holzzellen bei der gemeinen Kiefer (Pinus sylvestris). Jahrb. Wiss. Bot. 8: 401-420.

Saucier, J.R. & M.A. Taras. 1966. Wood density variation among six longleaf pine seed sources grown in Virginia. J. For. 64: 463-465.

Schreiner, E.J. 1935. Possibilities of improving pulping characteristics of pulpwoods by controlled hybridization of forest trees. Pap. Trade J. Techn. Sect. C: 105-109.

Shelbourne, C.J.A., B.J. Zobel & R.W. Stonecypher. 1969. The inheritance of compression wood and its genetic and phenotypic correlations with six other traits in five-year-old Loblolly pine. Silvae Genetica 18: 43-47.

Sijde, H.A. van der. 1979. Wood density assessment of selected trees from breast height samples. South Afr. For. J. 108: 42.

Sluder, E.R. 1972. Variation in specific gravity of yellow-poplar in the Southern Appalachians. Wood Science 5: 132-138.

Smith, W.J. 1967. The heritability of fibre characteristics and its application to wood quality improvement in forest trees. Silvae Genetica 16: 41-49.

Spears, J. 1979. Can the wet tropical forest survive? Commonw. For. Rev. 58: 165-180.

Spurr, S.H. & W.Y. Hsiung. 1954. Growth rate and specific gravity in conifers. J. For. 52: 191-200.

Spurr, S.H. & M.J. Hyvarinen. 1954. Wood fiber length as related to position in tree and growth. Bot. Rev. 20: 561-575.

Strickland, R.K. & R.E. Goddard. 1966. Correlation studies of slash pine tracheid length. For. Sci. 12: 54-62.

Taylor, F.W. 1968. Variations in the size and proportions of wood elements in yellow-poplar trees. Wood Sci. Technol. 2: 153-165.

Taylor, F.W. 1969. The effect of ray tissue on the specific gravity of wood. Wood and Fiber 1: 142-145.

Taylor, F.W. 1977. Variation in specific gravity and fiber length of selected hardwoods throughout the mid-south. For. Sci. 23: 190-194.

Teissier du Cros, E., J. Kleinschmit, P. Azoeuf & R. Hoslin. 1980. Spiral grain in beech, variability and heredity. Silvae Genetica 29: 5-13.

Theophrastus. 300 BC. Enquiry into plants. Transl. Sir Arthur Hart. 1966. Harvard Univ. Press, Cambridge.

Thor, E. 1964. Variation in Virginia pine. Part I. Natural variation in wood properties. J. For. 62: 258-262.

Wahlgren, H.E., D.R. Schumann, B.A. Bendtsen, R.L. Ethington & W.L. Galligan. 1975. Properties of major southern pines. I & II. U.S.D.A. For. Serv. Res. Pap. F.P.L. 176/177: 76 pp.

Webb, C.D. 1965. Variation in the wood of Sweetgum. Forest Farmer 24 (12): 9, 16-18.

Weiner, J. & L. Roth. 1966. The influence of environment and genetics on pulpwood quality. Inst. Paper Chem. Bibl. Ser. No. 224: 176 pp.

Westing, A.H. 1965. Formation and function of compression wood in gymnosperms. Bot. Rev. 31: 381-480.

Westing, A.H. 1968. Formation and function of compression wood in gymnosperms. II. Bot. Rev. 34: 51-78.

Winstead, J.E. 1967. Tracheid length as an ecotypic character in Liquidambar styraciflua L. under controlled conditions. Abstr. in Bull. Ecol. Soc. Amer. 48: 74.

Zobel, B. 1961. Inheritance of wood properties in Conifers. Silvae Genetica 10: 65-70.

Zobel, B. 1964. Breeding for wood properties in forest trees. Unasylva 18 (2-3), Nos. 73-74: 89-103.

Zobel, B. 1965a. Inheritance of spiral grain. In: Proc. Meet. Sect. 41, IUFRO, Melbourne, Vol. 1: 8 pp.

Zobel, B. 1965b. Inheritance of fiber characteristics and specific gravity in hardwoods – a review. Pap. Proc. IUFRO Sect. 41, CSIRO, Melbourne: 10 pp.

Zobel, B. 1970. Developing trees in the southeastern United States with wood qualities most desirable for paper. TAPPI 53: 2320-2325.

Zobel, B. 1971. Genetic manipulation of wood of southern pines including chemical characteristics. Wood Sci. Technol. 5: 255-271.

Zobel, B. 1976. Wood properties as affected by changes in the wood supply of southern pines. TAPPI 59: 126-128.

Zobel, B. 1980. Inherent differences affecting wood quality in fast-grown plantations. In: Proc. Symp. IUFRO Div. 5, Oxford, England: 169-188.

Zobel, B., J.B. Jett & R. Hutto. 1978. Improving wood density of short-rotation southern pine. TAPPI 61: 41-44.

Zobel, B., R.W. Stonecypher & C. Browne. 1968. Inheritance of spiral grain in young loblolly pine. For. Sci. 14: 376-379.

Zobel, B., R. Stonecypher, C. Browne & R.C. Kellison. 1966. Variation and inheritance of cellulose in the southern pines. TAPPI 49: 383-387.

Zobel, B., E. Thorbjornsen & F. Henson. 1960. Geographic, site and individual tree variation in wood properties of loblolly pine. Silvae Genetica 9: 149-158.

An anatomical explanation for visco-elastic and mechano-sorptive creep in wood, and effects of loading rate on strength

J.D. BOYD

Honorary Research Fellow, Division of Building Research, CSIRO, P.O. Box 56, Highett, Victoria 3190, Australia

Summary: For wood, it is widely known that the steady application of force, in either of the several possible stress modes, causes deformation which increases with time (creep). With any particular intensity of stress, it is well documented that the rate of creep varies significantly, according to whether the stress mode is in compression, bending, tension or shear. Additionally, it has been shown that the amount of creep tends to be considerably greater, if the moisture content of wood is reduced or cycled, than if it is constant (saturated or dry) during application of the force. However, the literature does not contain a broadly acceptable explanation for these phenomena.

It is demonstrated herein that creep generally, and also the qualitative differences in rates of response for the alternative stress modes, may be explained simply in terms of stress-induced physical interactions between the crystalline and non-crystalline components of the cell wall. An added influence of moisture reduction or moisture cycling is similarly explicable. Furthermore, it is shown that effects so deduced are fully compatible with the extensive experimental data in the literature.

Experimental data show that, after the forces causing creep are removed, much of the deformation of the wood is recoverable. The reasons for that become apparent from continuing interactions between the crystalline and non-crystalline wall components, which occur in response to removal of the initial actuating force. It is discussed how creep deflection, which is associated with axial compression and with bending, induces formation of microscopic crinkles across the general alignment of the microfibrils, which constitute the crystalline structural framework of the fibre wall. In turn, those crinkles predispose the fibres and the wood as a whole to failure at much lower intensities of stress, than can be sustained with force application over a short period.

Introduction

Wood and wood products are structural materials of great economic importance. The quality of their performance depends not only on strength, but also on their deformation or deflection under the various conditions of use. The amount of deflection varies widely in response to the duration of loading, and the associated temperature and moisture conditions of the environment. However, despite centuries of experience in such uses of wood, and despite the extensive advances in wood science during the last 50 years, the fundamental reasons for variations in structural performance are still unknown. This lack of knowledge may lead to serious inefficiencies in use.

Wood anatomists have given some attention to identifying the features of wood fibre (or tracheid*) walls, that exhibit evidence of the onset of failure, when wood is subjected to forces greater than it can sustain. That study is a matter of academic interest. However, it is infinitely more important to understand the anatomical cause of variable structural performance of wood, and especially the factors controlling the extent of its deformation, when subjected to forces over long periods in *normal service* conditions. Nevertheless, there is a very regrettable absence of such information in the literature.

It is hoped that the following discussion will lead to increased understanding of interactions of cell wall components, and stimulate efforts to obtain definitive anatomical data bearing on long-term performance of wood under stress. Particular attention will be given to the increase in deformation beyond the elastic deflection, *i.e.*, beyond the deflection which is immediately recoverable upon removal of the applied force. The added deformation is known as 'creep'. It varies widely with environmental conditions, and constitutes a flow within the fibre walls or between fibres. The creep deflection is partially recoverable over a period after removal of the applied forces.

Creep phenomena in wood and wood products are divided into two distinctive categories − visco-elastic and mechano-sorptive creep. *Visco-elastic creep* is the deformation beyond the immediate (elastically recoverable) deflection, that occurs with loading at constant moisture content of the material (green or dry). It increases with the duration of loading or force application and with the temperature level. Also it may involve effects of changes in temperature during loading. On the other hand, *mechano-sorptive creep* is that which occurs as a consequence of changes in moisture content of the material, while subjected to applied forces. Related strain phenomena and associated flow of material components occur if dimensional changes of specimens are limited or

---

* Herein, the term 'fibre' will be considered as representing either a fibre or a tracheid.

prevented during an extended period of force application, either with or without simultaneous moisture changes.

Mechanical testing of wood and wood products has yielded many data. Most of these are from 'static' tests, *i.e.,* with loading increased steadily to a maximum over a period of a few minutes; but others are with loads suddenly applied at maximum intensity (impact tests), and others again involve visco-elastic and mechano-sorptive conditions. Of the foregoing tests, most were made on bending and compression specimens; some others were on shear specimens, and relatively few were with tensile loading. Unfortunately there are practical difficulties in applying critical stresses in tests to determine tensile strength parameters parallel to the grain.

Recommended structural design procedures, for wood and wood products, include allowance for increased deflection due to creep effects. Similarly, a general provision is made for the fact that long-duration loading may lead to failures at stresses little more than half those which can be sustained in short-duration loading. However, these provisions for creep are rather crude and may be economically inefficient. At least partially, this is because the mechanics of the various creep effects within wood fibres and tissues are not understood. For example, it has not been known why a number of extensive moisture changes below *c.* 30 per cent moisture content, in a specimen subjected to a substantial applied force, may result in deformations more than five times those in similar specimens subjected to visco-elastic creep, and ten times more than the initial (elastic) effect, and may result in failure in quite a short period.

Theoretical models for creep effects

Schniewind (1968) reviewed the characteristics of the most significant models, that had been proposed to explain creep in wood and wood products, during the exceptionally active research period of 10 years prior to that date. Most of those theoretical models were intended to represent visco-elastic creep effects, and they involved claims that responses could be represented by linear relationships. However, Schniewind noted that, although some data appeared to show linear associations in some conditions and over very restricted ranges of stress and strain effects, the linearity assumption was invalid generally. In any case, those theories were completely inadequate in respect of mechano-sorptive effects. Grossman (1976) noted additional theories that had been proposed subsequently to 1968, particularly to represent mechano-sorptive creep responses. However, he commented that the data showed those hypotheses were unsatisfactory.

Necessary characteristics of an adequate model

Grossman (1976) outlined the various features of mechano-sorptive behaviour which had been recorded. He suggested that for wood and wood products, an acceptable model should: (1) indicate reasons for the typical mechano-sorptive reactions; (2) be closely related to the basic structure of wood; and (3) be compatible with all the data, including those exhibiting responses to previous stress history of the specimens. In the study herein, it is contended that such a model should be completely compatible also with both elastic and visco-elastic reactions of wood, wood fibres and wood products. Particularly also, it should indicate how mechano-sorptive creep can lead to failure at much lower applied stresses, or with much shorter duration of loading than would be the situation in the absence of the moisture changes.

Furthermore, the model should be compatible with the speed of application of loading having a significant effect on the relative strength shown by specimens subjected to either dynamic, static or long-duration loading. On the other hand, the model should indicate why the mechano-sorptive creep is determined by the extent of the moisture change and not by the rate at which that change occurs. Finally, the model should facilitate an explanation of the strain recovery which is associated with visco-elastic and mechano-sorptive creep. Herein, prime consideration will be given to describing a model which appears satisfactory to explain visco-elastic creep, and associated data; and its adaptability to explain mechano-sorptive creep will be demonstrated.

Anatomical determinants of rheological responses of wood

A hypothesis will be proposed to relate certain dominant features of the anatomy of wood fibres, and a biophysical factor associated with their differentiation, to the reactions of wood tissue to applied forces, under the variety of conditions of loading referred to above. Subsequently, the adequacy of the theory will be tested against the published data for the various relevant studies.

An impression of the general nature of wood tissues is a necessary background for effective examination of this subject area. For simplification however, the following discussion will be centred on the fibres because: (1) they are all similarly oriented; (2) they greatly outnumber all other cell forms; and (3) the thickness of their walls, and therefore their rigidity and strength are the predominant determinants of the responses of the tissue to applied forces. Ultimately, those structural responses must be related directly to the components and form of the fibre wall.

In a broad way, the structural responses of the fibre wall are determined by a somewhat complex framework of crystalline cellulosic microfibrils, and inter-penetrating non-crystalline substances. A knowledge of the physical attributes of those wall components is essential to an appreciation of their responses.

Significant features of microfibril arrangements

Observations on many electron micrographs show that, except in the structurally-weak primary wall, adjacent microfibrils follow generally similar, parallel helical paths within lamellae; and a large number of such lamellae develop in concentric, adjacent positions to constitute the structural framework of the cell wall. If those walls are treated to remove the non-crystalline material, it becomes apparent that bonding exists between microfibrils, at randomly-spaced positions along their length (Boyd and Foster, 1975).

Partly due to the initial lack of absolute straightness as microfibrils are formed (Boyd, 1980b, 1982*) and partly due to a swelling of the amorphous matrix material which fills the spaces between adjacent bonded positions on the microfibrils (Boyd, 1972), those basic structural units develop configurations which have been described as lenticular trellis forms (Boyd and Foster, 1975). Generally the widths of the trellis openings are small relative to distances between bonded positions along the length of the microfibrils, and it is apparent that the average width increases in proportion to the departure of the mean microfibril orientation relative to the axial direction of the fibre (Fig. 1).

In a study of the distribution of lignin in the fibre wall, Scallan (1974) concluded that: (1) the 'lignin-hemicellulose gel was located preferably in tangential arrangement', but 'randomly' and not in the form of a lamella; and (2) it occupied 'cleavages (between microfibrils) in the radial direction'. He illustrated his interpretation of the positions of lignin and cellulose with two alternative configurations of the microfibrils. The front face (vertical) of Fig. 2 represents his suggested form of a radial face; also, the left side of the (transverse) face accords with his proposal. The right side of the transverse face shows an arrangement more directly comparable to that on the radial face. The right side is based on the conclusions of Boyd and Foster (1975). These led to the thought that the curvatures of microfibrils, around the lenticular openings, are likely to involve some local relative side displacements of a large proportion of the adjacent microfibrils, rather than an occasional large opening as shown in Scallan's sketch.

---

* It is anticipated that a more comprehensive paper, 'Biophysics of microfibril orientation in plant cell walls', will be published elsewhere in 1982.

176

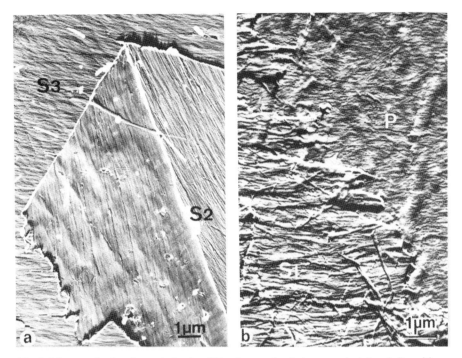

Fig. 1. Micrographs showing variation in width of separation between contact (bonded) positions along microfibrils. (a) Compare usual narrow openings between axial microfibrils in S_2, with the wider openings in S_3. (b) Openings in S_1 are similar to those in S_3 (disregard disturbed orientations of a few microfibrils at the surface). Micrographs, of *Picea jezoensis* Carr, by courtesy of Prof. H. Harada, Department of Wood Science and Technology, Kyoto University, Japan.

Kerr and Goring (1975) proposed a model in which the appearance of the masses of lignin-hemicellulose gel, within the microfibrillar framework, was somewhat similar to Scallan's, though with rectangular boundaries on the transverse face, and apparently without change of form (*i.e.*, as continuous axial 'strips') along the length of the fibres. They also noted that Scallan's suggested form, of lignin and matrix accumulations, constituted 'disc-shaped, tangentially oriented platelets', and showed similarities to their 'interrupted lamella structure'.

Fig. 3 represents the form of structure proposed by Boyd and Foster (1975). Although determined independently, it is very similar to that proposed by Scallan, and it also would lead to lignin-hemicellulose (matrix) gel being distributed mainly as lens-shaped platelets, However, because of the random locations of the microfibril to microfibril bonding, bridges would link the platelets comprehensively in both the radial and tangential directions. Accordingly, if the cel-

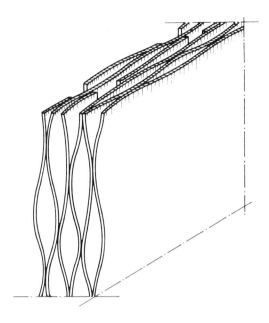

Fig. 2. Concept of arrangements of microfibrils, as they enclose lignin and matrix materials in lenticular platelet forms in S_2; the width of openings is exaggerated for clarity of presentations. The left side of the tangential (long) face illustrates Scallan's (1974) concept; the right side represents the concept of Boyd and Foster (1975).

lulose, hemicellulose and pectin were dissolved out of the cell walls, a coherent lignin skeleton should remain; such a 'structural' form was demonstrated by Bentum *et al.* (1969) and Harada (1965).

It has been suggested and supported with biophysical argument, that very large stresses in trees are generated in the walls of wood fibres during their differentiation (Boyd, 1972). Extensive data such as those of Jacobs (1938), Boyd (1950, 1977a, 1980a), Nicholson (1973), and Boyd and Foster (1974) provide strong though indirect support for that conclusion. Other substantial data and argument (Boyd, 1982) indicate that: (1) strains associated with stress development, during the formation of microfibrils within the cell wall, determine the general or mean orientation of those microfibrils; and (2) biophysical conditions lead to frequent small variations of microfibril orientation relative to their mean direction; in turn, that leads to the development of 'chains' of small lens-like openings between randomly-spaced positions of inter-fibril bonding.

One reason advanced, to explain the lenticular trellis arrangements, was that the microfibrils cannot be formed in a state of tension, such as would be neces-

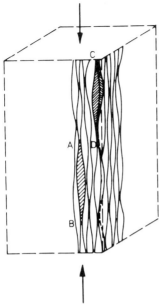

Fig. 3. Microfibril and matrix forms as discussed by Boyd and Foster (1975). The shaded form 'AB' represents the matrix enclosed between microfibrils as seen on one face; 'CD' is a three-dimensional view, as at the intersection of radial and tangential faces.

sary to prevent local variations of microfibril direction relative to their mean orientation (Boyd, 1982). Another conclusion of that study, was that associated with the extension of the microfibrils by tip growth (Preston, 1974), the stimulus for both their formation and their basic orientation must be related to the orientation of the most severely strained microfibrils in the cell walls. As those microfibrils are in lamellae well below the inner face of the cell wall, then the stimulus for continued and controlled microfibril formation must be conveyed to the plasmalemma through the several overlying, non-strained lamellae of microfibrils. Contact between the plasmalemma and the overlying 'blanket' of microfibrils, which lie in directions 'meandering' somewhat around their mean orientations, must tend to cloud the precise 'image' of optimum orientation for differentiation of new reinforcing microfibrils. Ideally, the latter alignment would be indicated to the plasmalemma, by the highly-strained microfibrils more distant from the inner face of the wall (Boyd, 1982).

Furthermore, subsequent to the formation of microfibrils in a series of slight variations around their mean direction, there are likely to be some modifications of those initial divergences (Boyd, 1982). That is due to the powerful axial stress system which develops in the fibre wall, and to its interaction with forces in the older wood tissue to which the fibre is attached (Boyd, 1950, 1972). Thus during differentiation of a particular cell wall, there is a tendency for axial tension to draw out microfibrils that were formed in near-axial orientations. Accordingly, they would tend to become more straight and parallel, and correspondingly the width of the lens-like openings along their length would be reduced somewhat. On the other hand, axial tension would tend to cause an increase in the initial transverse width of openings between microfibrils formed in near-transverse orientations.

This effect has been observed in cells of other types of plants. For example, Frei and Preston (1961) noted the effect in growing cell walls of algae. Where such stress-stimulated modifications of microfibril arrangements occur in wood fibres, they could account for the general differences between widths of lenticular openings, such as occur between microfibrils in the S_1 layer relative to those in the S_2 layer, and between those in the S_2 and S_3 layers of the cell wall (Fig. 1).

The nature of the non-crystalline components of the cell wall

Research papers and reviews, such as that of Wardrop (1971), indicate that it has long been established that the spaces between the microfibrils in the structural framework of the wood cell wall are filled with matrix materials and lignin. The matrix components consist essentially of non-cellulosic polysaccharides and their derivatives. Wardrop noted that the matrix and lignin components were predominantly amorphous in form. Frey-Wyssling (1976) made the significant statement that − 'the matrix is a hydrophilic gel which by its high swelling capacity and its plastic deformability is in a state between solid and liquid', and has a 'high water content'.

It is well documented also that during lignification, plant cell walls show an associated swelling. Qualitative observations of that phenomenon have been noted by Frey (1926), Alexandrov and Djaparidze (1927), Preston (1941), Preston and Middlebrook (1949), and Onaka (1949). Analyses by Boyd (1972), of measurements made by Grozdits and Ifju (1969) on differentiating tracheids of eastern hemlock, showed that the cells expanded more than 20 per cent in the radial direction during lignification.

Furthermore, it has been shown that there is a highly significant correlation between volumetric shrinkage of wood and lignin content (Kelsey, 1963). Thus the lignin acts as a bulking agent in the cell wall, and tends to resist a reduction in the dimensions of the wall, the cell and the tissue. That characteristic has been confirmed by comparing the shrinkage of tissues before and after removal of a portion of the lignin (Choong, 1969). The presence of wood extractives within the cell wall has a similar bulking effect (Nearn, 1955; Choong, 1969). Apparently the extractives fill spaces or pores in the cell wall, that could otherwise be filled with water (Stamm, 1964). That has been clearly indicated by the fact that the presence of extractives is associated with a low equilibrium moisture content for the wood, whereas their removal raises that level (Nearn, 1955; Stamm, 1964).

Direct effects of moisture within the cell wall

In reference to free swelling and shrinking effects in general, Stamm (1964) stated: (1) 'the absorbent must be a plastic solid'; (2) 'the absorbate must have sufficient affinity for the absorbent to spontaneously form intimate solid solutions with the absorbent, accompanied by evolution of heat'; and he also stated (3) 'the wood-water and cellulose-water systems meet both of these requirements'.

Stamm remarked that the only position where water is absorbed by cellulose materials is in the amorphous region of microfibrils. He suggested that it is absorbed there only between the long cellulose molecular chains, which are thought to be far from perfectly oriented. He suggested also that, when the water evaporates on desorption, surface tension forces are set up over the area of contact of the water with the cellulose chains, and that tends to draw the chains together.

In respect of the amorphous gel, Stamm stated that the effect of structural voids on the 'shrinkage of isotropic gels is entirely inward'. Correspondingly, the 'swelling of isotropic gels is entirely outwards'. He noted that this is not the same as the effect on the fibre as a whole. His explanation is that the stiffness of the transversely oriented microfibrils in the S_3 layer of the fibre prevents shrinkage of the cell lumen, and correspondingly, the transverse arrangement of microfibrils in S_1 restricts an outwards swelling tendency of the wall.

Referring to the effect of frequent loading on failure of wood specimens, Stamm noted that the number of stress-strain cycles that a material will endure before failure is a measure of its 'fatigue life or endurance'. Also, 'below a definite stress level, frequently failure does not occur even after millions of cycles'. In addition, 'moisture and plasticizers that reduce the brittleness of fibres greatly increase their fatigue life'. On the other hand, 'the smoothness of the surface (of the material) and the presence of surface flaws affect (reduce) greatly the number of cycles of tension, bending, or torsion that a specimen will endure'.

In respect of any particular load application, Stamm noted the well-known

Table 1. Young's Modulus of Elasticity Values* (MPa) $\times 10^3$

	Matrix Material		Microfibril Framework		
Estimate	Axial and Transverse	Estimate	Axial (A) Direction	Transverse (T) Direction	Ratio A/T
Best	2.0	Case 1	319	37.2	8.6
Very stiff	6.9	Case 2	246	16.4	15.0
Very compliant	0.2	Case 3	134	27.2.	4.9

* Data are from Mark and Gillis (1973)

Table 2. Ratio of Young's Modulus to Shear Modulus*

	Matrix Material		Microfibril Framework	
Estimate	$\frac{E}{G}$	Estimate	$\frac{E(axial)}{G}$	$\frac{E(transverse)}{G}$
Best	2.6	Case 1	816	95
Very stiff	2.6	Case 2	1453	97
Very compliant	2.6	Case 3	31	9

* Data are form Mark and Gillis (1973)
'E' represents Young's modulus of elasticity, and 'G' represents 'shear modulus of elasticity'

fact that deflection increases with a rise in temperature. Additionally, green (non-seasoned) wood exhibits greater creep deformation than dry wood. Similarly for moisture contents below 'saturation' of the fibre wall, it has been reported that generally the higher the moisture content the greater the creep.

Relative rigidity of microfibril and matrix components

Mark and Gillis (1973) made a broad theoretical study of the relative stiffness of the microfibrillar framework and the matrix material in the wall of the wood fibre. Their basic estimates of the comparative rigidity of those components, in respect of direct stresses (axial and transverse relative to the axial direction of the fibre), have been assembled here in Table 1. Additionally their values for ratios of rigidity under both axial and transverse stresses, relative to the corresponding rigidities in shear, are reproduced in Table 2.

In respect of the values for the cellulosic framework in both those tables, the authors provided three estimates based on different approaches; those estimates could be regarded as covering upper and lower extreme limits of conceivable values, and an intermediate one. Relative to the matrix material, they stated that the range of variation 'encompasses all reasonable changes in properties..... due to moisture content variation'. They regarded their 'very compliant' value as representative of the unseasoned condition of the fibre, and their 'best' estimate as representative of the typical seasoned condition; the 'very stiff' estimate was considered a somewhat unlikely extreme.

Mark and Gillis remarked that the 'matrix properties are highly sensitive to moisture changes, whereas the cellulose I lattice (basic crystalline structure of the elementary fibrils, which together constitute the microfibrils) is impervious to water penetration. Thus, the changes in the fibre properties with moisture are the result of molecular mobility within the matrix and at the interfacial surface of the matrix and the cellulosic microfibrils'. Their analysis, of the effects of different microfibril angles in the S_2 layer, showed that this conclusion was valid for the complete range of likely variations of both microfibril angle and moisture content.

Strain interactions between microfibrillar framework and matrix material

Overall, the data above indicate that, in whatever conditions a specimen is subjected to an applied force, so as to cause bending, tension, compression, or shear, the dominating rigidity of the microfibrils will determine the initial (elastic) deflection or strain of the cell and the tissue. At the same time, because the matrix material is contained within the microfibrillar framework, in arrangements such as illustrated in Fig. 3, the matrix will be forced to adjust its position and shape to maintain physical compatibility. Thus it must conform with the modifications to the alignments of the microfibrils, that are imposed by the applied force. Considering the matrix as a hydrophilic gel (Frey-Wyssling, 1976), which exhibits a degree of bonding to elements within itself and to the microfibrils, the adjustment to its form (location), that is imposed by the microfibrils, requires that it must 'flow' into the newly determined containment limits. Undoubtedly the extent of that flow will be limited, but its nature must be in accord with the well-known characteristics of viscous flow.

Viscous flow generates shear forces within the matrix, and frictional resistance to flow develops along its boundaries with the microfibrils. Also, the forced relocation of the matrix involves some breaking of bonds. The resistance to such adjustments to its location would be least when the matrix was saturated with water, and with increased temperatures, as both factors minimise the viscosity of a gel (Eshbach, 1952; Stamm, 1964). Where some matrix material has been forced to flow to a new location, the corresponding loss of its constraint, on the deflection of the microfibrils, would allow the externally-applied force to increase the deflection. In turn, that would cause more flow of the matrix. With some delay arising from the slow viscous flow of the matrix, that creep mechanism would proceed, albeit at a reducing rate, until there was a nominal balance between the force which the deflecting mircofibrils impose on the matrix, and the resistance of the matrix to continued flow. Theoretically the stage of absolute balance would never be reached, but the strain rate tends to become infinitesimally slow, unless the specimen is close to failure.

Basic form of the anatomical model

The most significant characteristic of the anatomical model, as a representation of the deflection responses of wood fibres, to different types of load application in all environmental conditions, is that it is based on the lenticular trellis arrangements of the microfibrils. As may be observed in Fig. 1 for wood, and as previously illustrated and discussed in some detail (Boyd and Foster, 1975), that lenticular microfibrillar arrangement appears to be general throughout the range of aquatic and terrestrial plants, both during extension growth of the cells, and during secondary wall thickening. About the same time as the study cited above, and from a different direction of investigation, Scallan (1974) proposed a comparable model for the arrangement of microfibrils and the associated inclusion of matrix materials. Also Boyd (1982) defined and discussed the biophysical factors which lead to the microfibrils developing the lenticular trellis form initially.

The orientation of microfibrils, between pairs of bonded positions along the length of adjacent ones, diverges slightly from the mean direction and then converges back again. With an exaggeration of the width of the separations, to facilitate appreciation of arguments in the following discussion, Fig. 3 illustrates an idealised (abnormally regular) but exaggerated form of the lenticular configurations. Considering the block of lamellae as cut from the cell wall, it will be appreciated how similarly-directed microfibrils in an overlying lamella would tend to rise above the face as they crossed over parts of the underlying microfibrils. They would then be forced to dip slightly into the alternating lens-like openings, as they extended under the influence of the turgor pressure in the cell. Accordingly, in electron micrographs the appearance of the surface may convey the impression of some inter-weaving of microfibrils.

That formation process would account for lenticular forms developing in

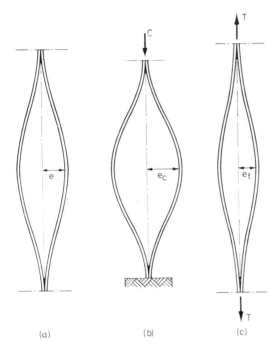

Fig. 4. Effect of axial stresses on microfibril configurations between bonded positions along their length. (a) Schematic arrangement in non-stressed form, with a departure 'e' from the mean direction of the microfibrils. (b) Compression by a force such as 'C', causes an increased departure 'e_c', relative to the mean direction. (c) Tensile forces 'T' cause a reduction in microfibril departure to 'e_t'.

both tangential and radial planes within the fibre (and in respect of the axial direction in the tree). Accordingly the lens-like forms, as illustrated between A and B (Fig. 3) would appear as boat-like shapes (three-dimensionally), as between C and D (Fig. 3). Alternatively, the traditional weaving shuttle may provide the best representation of such forms enclosed between both adjacent microfibrils within a lamella and those in adjacent lamellae. Such outlines would determine the shape of matrix accumulations throughout the cell wall.

Basic responses of the anatomical model to stress

Because the microfibrillar framework is very much stiffer than the matrix material (as discussed before), when a force is applied to the fibre wall, the microfibrils will carry virtually all of the applied load. That reaction is in accordance with a well-known law of elasticity, which states that the load carried by each structural element, in a combination of elements of differing rigidity to which the load is applied, will be in proportion to the relative stiffness of each element (*e.g.* Timoshenko, 1940). If the microfibrils have a curved form, such as between A and B in Fig. 3, and the applied force is an axial compressive one, then a similar shape between bonded positions, such as in Fig. 4a, will deflect under a compressive force 'C' to a position such as in Fig. 4b. Accordingly, the width of opening between pairs of such microfibrils would be increased; and the axial length, over which their curved form extended initially, would be reduced. Correspondingly, if a tensile force 'T' were applied to pairs of micro-fibrils with a form as in Fig. 4a, the width of openings would be reduced, and the axial extent of their length would be increased (Fig. 4c). The changes in the outlines of the microfibrils would force compatible changes of shape on the matrix material contained in the lenticular configurations between bonded positions, such as from Fig. 4a to Fig. 4b or Fig. 4c.

If then the externally applied force were removed, the resilience of the crystalline microfibrils, in association with the elastic strain energy stored in them as a consequence of their deflection by the applied load, would initiate a reaction tendency for them to spring back to their original configurations. That reaction would tend to force the contained matrix material back towards its original position and form. Additionally, relief of the stresses in the microfibrils, that resulted from the externally-applied force, would lead to a stress reaction (recovery) arising from the strain energy stored through compression of the matrix. However, the frictional forces associated with both the initial and subsequent movements of matrix material, would necessarily cause dissipation of some energy, in the form of heat. Hence the original configurations of microfibrils and matrix could not be fully restored, and there would be a resulting permanent deformation (flow or residual creep effect).

On the other hand, if for example the original forced deformation were a result of an applied compressive force, and if at that time the matrix material were swollen to its maximum volume by contained water, then a reduction of moisture content while under load would lead to a different situation. The matrix would tend to shrink away from the microfibrils which determined its outer boundaries (Stamm, 1964), so that in all adjacent lens-like configurations, as illustrated in Fig. 3 and 4, the effect would be as illustrated in Fig. 5. That would remove the previous constraint of the matrix on the amount of microfibril deflection under the applied load. In consequence, there would be a corresponding increased deflection of the microfibrils.

If later, with the compressive force still active, there were an increase in the moisture content, that would swell the matrix and tend to force the microfibrils

186

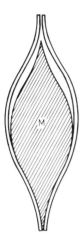

Fig. 5. Effect of loss of water on matrix material 'M' between microfibrils.

back to their configuration prior to the loss of moisture. Also, assuming that the lateral deflection of the microfibrils was less than the critical buckling condition for such elastic columns* supporting the compressive force 'C', there would be an increment of strain energy stored in them, as a consequence of the increase in deflection associated with the earlier shrinking of the matrix. Accordingly, the resiliance of the microfibrils would facilitate a return of the previously-stored strain energy increment. As a result, they would tend to regain their earlier configuration. In turn, the movement of the microfibrils would stimulate flow of the matrix back towards its configuration (and location) before the desorption phase.

However, during that cycle of desorption, some energy would have been dissipated in the breaking and remaking of bonds, and some would have been dissipated as heat caused by friction associated with viscous flow of the matrix. Consequently, the deflection which occurred directly as a result of loading associated with desorption could not be fully recovered during sorption; *i.e.,* a moisture cycle under load would cause a net increase in the residual creep.

It will be discussed later herein, how permanent crinkles develop gradually in fibre walls, and particularly in the microfibrils which constitute the S_2 layer. Furthermore, that crinkling is initiated at intensities of compressive stress that

* Except during development of incipient failure modes, as will be discussed later, critical buckling would generally be prevented because of interactions of microfibrils in adjacent positions throughout the S_2 layer, and because of the overall restraint on lateral bulging of S_2, which is imposed by the transversely oriented microfibrils in layers S_1 and S_3.

are relatively low in relation to failure loads in testing over short periods. It will be shown also that the number and severity of those crinkles, throughout the length of the fibre, would increase with moisture cycling under load, and with the duration of loading. Together, these effects must lead to increasing deformation or creep of the fibres, and of wood tissue as a composite. Creep data show that such deformation is substantially recoverable with moisture cycling after removal of the load. Generally however, other data show that the crinkles in the microfibrils can lead to significant loss of strength and to increasing liability to early failure under load.

Relationships between data on creep and anatomical responses

With reference to the wide range of experimental work on visco-elastic creep, and more particularly that on mechano-sorptive behaviour of wood, it is proposed to show how the nature of data on the responses of the material may be explained by the reactions of the main components of the walls of wood fibres. To minimise listing of all references to categories of creep data, generally appropriate reviews will be cited, and also characteristic references which may quote others of relevance.

Visco-elastic creep at constant temperature

This involves the creep strain or deflection due to an applied force, that is induced without a significant change in either the moisture content or the temperature of the specimens. Research reviewed by Schniewind (1968) showed that visco-elastic creep increases with both the duration and intensity of loading. Also, data show that the creep response increases with both higher (constant) temperature and higher (constant) moisture content of the fibre wall, within the moisture range from dry to saturation level in the wall. Furthermore, it has been shown that the amount of creep varies with the type of stress imposed, as between compression, bending, shear and tension. It is convenient to discuss these general factor interactions initially in respect of axial compressive loading. The variations in responses with other forms of applied stress will be outlined later.

Basic form of the deflection response. − The discussion of the interaction between the microfibrillar framework and the gel-like matrix demonstrated that an axial compressive force would tend to cause an increase in the width of lenticular openings. Simultaneously, pressure from the deflecting microfibrils would cause a viscous flow of the enveloped matrix material, towards conformity with a modified, stable, enclosing configuration of the microfibrils. In the previous section, the physical reactions associated with the change of alignment of the microfibrils were outlined, and also the physical factors associated with the consequent viscous flow of the matrix. Because the initial position and rigidity of the matrix material imposes a restraint on the extent of the deflection of the microfibrils caused by the applied force (as discussed before), the time involved in the slow viscous flow of the matrix, towards a new conforming configuration, constitutes a critical factor in the strain response of the cell wall.

Thus the elastic response tendency of the relatively rigid microfibrillar framework is modified, by matrix interaction, into a response continued over time, *i.e.*, to a creep response. Immediately after the application of the load, there must be the maximum out-of-balance in respect of developing a stabilising reaction of the microfibrils to limit deflection within the cell wall. The corresponding initial pressure on the matrix, causing flow to a conformable position, must also be a maximum. As viscous flow progresses, in the process of reaching conformability of the matrix in a stable configuration of microfibrils and matrix, the size of the reactive force impelling the viscous flow must fall off at an exponential rate. Accordingly, the natural responses of both the microfibrils and matrix, to any form of loading of the wood fibre wall (and of bulk wood) must induce the typical visco-elastic creep responses which have been recorded, and also described mathematically by many researchers.

Influence of load intensity. – An increase in the intensity of loading does not introduce any change in the basic form of the interaction between microfibrils and matrix, during cell wall reactions to counterbalance the effect of the applied force. Accordingly, the form of creep reaction would generally be similar, although the amount of creep would increase with increasing load intensity. Published experimental data, such as those of Kingston (1962), have indicated this.

In a paper being prepared for publication, Nakai and Grossman* have shown this effect of load intensity in bending tests, with maximum stress ranging from 17 per cent to 67 per cent of the estimated maximum strength in static loading. There is no reason to doubt that creep behaviour is similar down to the lowest proportional stresses at which creep can be measured accurately. For example, Kingston and Armstrong (1951) and Kingston (1962) noted that irrecoverable creep in bending occurred at stresses as little as 3 to 4 per cent of the ultimate stress. In tension tests parallel to the fibre, Murphy (1963) used

---

* Based on research results obtained at the Division of Building Research, CSIRO.

X-ray diffraction to demonstrate that elastic reactions of the cell wall were similar, although progressively larger at stresses of 12, 25, 50 and 100 per cent respectively, of the estimated ultimate failure load. He demonstrated also that there were creep effects with similar relationships. Additionally, in tests with various levels of force restraining shrinkage perpendicular to the grain, during the drying of wood of four species, Fujita and Takahashi (1969) showed that the creep in tension had similar proportional effects on the mechanical properties. Previously discussed physical reactions of the microfibrils and matrix materials are compatible with such responses.

Influence of moisture content and temperature. – As pointed out by Stamm (1964), increases in moisture content up to saturation level in the cell wall, and also increases in temperature have the effect of plasticisers. Accordingly, both would cause a reduction in viscosity and lowered resistance of the matrix to flow. Hence, both factors would increase the rate of creep deflection under load. That such is the response in practice is borne out by many data (Schniewind, 1968). Thus, the effects of those factors are compatible with the general thesis, that the creep induced by a force applied over a given period is determined essentially by the interaction between the induced deflection of the microfibrillar framework, and the resistance to viscous flow that is offered by the matrix material.

Comparative responses to compressive and tensile stresses. – Independently of the nature or direction of the force, the same primary interaction responses are involved between deflecting microfibrils, and a viscous matrix material which must conform to their shape adjustments. Depending on strain direction however, there are secondary factors which influence the practical extent of adjustments to the outlines of both of these basic wall components.

For example, mirofibrils in forms such as in Fig. 4a, when subjected to compressive force as in Fig. 4b, develop an increased eccentricity 'e_c' of the axial loading relative to their deflected middle section. Thus the curvatures of such microfibrils approach more circular forms. Hence the volumes enclosed between pairs or groups of adjacent ones approach the maximum possible amount; the latter is associated with a spherical form. However, the practical extent of such a shape change is severely limited, because of simultaneous interactions of similar effects in adjacent lenticular forms. Nevertheless, the matrix material becomes more readily accommodated, and its capacity to impose constraint on further deflection of the microfibrils is correspondingly reduced.

Under tensile forces on the other hand, the microfibrils tend to approach a straight and parallel form, as shown by the comparison of Fig. 4a and Fig. 4c. Hence there is a reduced eccentricity (e_t) of the applied force; and the bending stress on the microfibrils, that is associated with the direct stress imposed by the tensile force, is progressively reduced. Correspondingly, the maximum stress and the potential for extension (strain) of the microfibrils is reduced (or their relative rigidities are increased) between pairs of bonded positions along their length.

Thus with tensile stress, the volume contained between adjacent microfibrils in the lenticular trellis configuration is reduced progressively. Accordingly, it must become increasingly difficult to force the matrix material to occupy the reducing space available between the microfibrils. The study by Simpson (1971) indicated that a small reduction in the volume of the matrix gel could be induced by axial tensile strains on the microfibrils. That involves a reduction in pore spaces and moisture content, as a consequence of compressing the matrix into the reduced space. Hence with cell walls in which the mean direction of the microfibrils in the S_2 approaches axial orientation, as in normal wood and tension wood fibres, it is inevitable that creep induced by a tensile stress will be much less than would be induced by a similar compressive stress.

Effect of bending stresses. – When bending stresses are imposed on a beam, the resulting deflected form (Fig. 6a) involves the development of compressive stresses on the concave side and tensile stresses on the convex side of the beam. If the associated strains were fully elastic, the stress distribution would take the form indicated in Fig. 6b. This shows a uniform gradation in the intensity of both tensile and compressive stresses (and corresponding strains), from a maximum at the outer face to zero at the centre of depth of the beam. In compression however, the extent of a nominally elastic relationship between stress and strain is much less than in tension. Kollman and Côté (1968) illustrated that by the stress/strain relationships to failure, as shown in Fig. 6c. The consequences of that relationship are in accord with the responses which would be expected on the basis of microfibril and matrix configurations discussed above, and the reactions as discussed below.

Because of the increasing creep associated with increasing compressive stress, the most highly stressed outer fibres on the compression side of the beam will yield (deform) more per unit of applied bending stress on the beam than adjacent ones towards the middle of its depth. In that yielding condition, the fibres are less rigid than adjacent less-strained ones. That will have the effect of transferring part of the force originally on the outer fibres to those on their inner side. In turn, they will creep and yield more. This will happen progressively to fibres more distant from the compression face of the beam. With

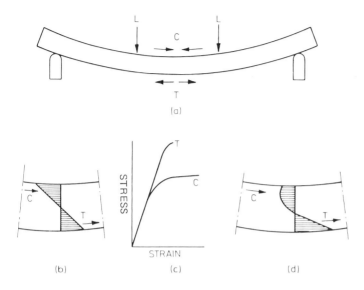

Fig. 6. Stress distribution in beams. (a) Loads 'L' cause compressive stresses 'C' on the upper side and tensile stresses 'T' on the lower side. (b) Straight-line variations of strain intensities within the elastic range, from maximum compressive strain at the upper face, to maximum tensile strain at the lower face. (c) Variations from elastic (straight line) stress-strain relationships to ultimate failure stress, in tension 'T', and compression 'C'. (d) Modified strain relationships in beams, due to increasing load duration and strain.

increasing bending stress, the result will be to modify the stress distribution on fibres subjected to compression, from the form in Fig. 6b to that in Fig. 6d.

On the other hand, the rate of creep in tension is much less than that of compression. Furthermore, it appears from previous discussion of comparative responses to stress, that any tensile creep tendency would have the effect of making the fibres increasingly elastic, except at stresses very close to failure levels. Consequently, the stress distribution in the tensile zone in the beam would retain a virtual straight-line relationship (Fig. 6d). However the form of compressive stress distribution, and the stability requirement that the bending moment and tensile and compressive reactions be in balance, leads to a lowering of the neutral axis (zero stress position) in the beam. Accordingly the intensities of the tensile stresses, at all distances from the neutral axis, would be increased in the process of maintaining a moment of resistance which counterbalances the applied bending moment, and stabilises (nominally) the form of the deflected beam. There would be a corresponding increase (small) in deflection due to the tensile straining. However, because of creep, the deflection of the beam can never be completely stabilised.

It is obvious from the upper part of Fig. 6d, that the distribution of compressive stresses in this part of the beam approaches that for a specimen subjected to axial compressive stresses alone. However, because the lower part of the beam is subjected to tensile stresses only, less creep is induced there. The creep reaction of the beam as a whole must be a composite of the two dissimilar creep responses. Hence the deflection of the beam, due to visco-elastic creep, will be somewhere intermediate between that caused by the relatively large creep response to axial compression on the upper side, and the effect on the lower side of the much smaller creep response to axial tension. Relevant to that, Kingston and Armstrong (1951) noted that 'a considerably greater change in length was found to occur in the compression than in the tension face' of the beam. The relative deformation ratio 'was 3 to 4:1'.

In a beam subjected to forces as in Fig. 6a, the counterbalancing induced reaction involves vertical and horizontal shear stresses, in addition to the compressive and tensile stresses discussed above. However, standard texts on elasticity of materials (*e.g.* Timoshenko, 1940) show that the usual deflection due to shear stress is likely to be not more than a few per cent of that due to tensile and compressive stresses. Correspondingly, the contribution of the shear stresses to the total creep of the beam would be comparatively small.

Effect of shear stresses alone. – It is difficult to relate anatomical responses directly to creep in shear. Largely this is because 'pure' shear stressing rarely, if ever, develops a simple shear plane effect in wood elements. Relative to investigation using many different types of shear testing apparatus, Kollmann and Côté (1968) stated that problems arise because bending stresses are inadvertently imposed, and other factors complicate the stress reactions in tests to study the shear phenomenon. As outlined above, bending stresses cause some fibres to be stressed in compression, some in tension, and all in shear, with a range of intensities of each strain type. Furthermore, as pointed out in texts such as that of Timoshenko (1940), shear stresses in one plane are invariably associated with shear stresses in a plane at right angles, and also 'pure shear is equivalent to the state of stress produced by tension in one direction and an equal compression in the perpendicular direction'.

The general structure of wood might be represented by a large number of axially-oriented fibres which are 'glued' to each other through their bonding to intermediate middle lamellae zones of matrix material, that surround the full periphery of each fibre. Discussion earlier, of the relative rigidities of the matrix material (as in the middle lamella region) and the microfibrillar framework, indicated that the intensity of force, which is needed to develop significant shear stress and viscous flow (creep) of the matrix, would be much smaller than that required to cause a comparable strain in the microfibrillar framework. However, boundaries between adjacent fibres never lie completely on a single shear plane. That situation arises largely because the tips of fibres extend to penetrate between other fibres, while they develop variations of conical shapes to fill the available spaces.

Accordingly, no single shear plane could pass through intercellular zones only for a single fibre, and certainly not for a number of fibres. Accordingly there would be a transfer of the associated strain through cell walls, and hence to fibres distant from that plane. In turn, that would involve effects of both tensile and compressive reactions (and strains) in those fibres, and also effects of associated shear stresses at right angles. Thus reactions comparable to those in a loaded beam must be developed in fibres subjected to the influence of the shear force.

Accordingly the creep response of wood, to the application of a shearing force, is an extremely complex one. It could be expected that at a given stress, the creep deformation in shear would be considerably greater than that for pure tension. However, its severity cannot be determined readily from theoretical analyses, and there is a dearth of experimental data to indicate the precise situation. Possibly the severity of the creep in shear may be somewhere between that when uniform compression is applied, and that in bending. At least there are some unpublished data by Grossman and Kingston*, which show that the amount of creep in shear increases in proportion to increases in temperature and moisture content, of otherwise similar test specimens. That result is compatible with data for other types of stress application, and it indicates that the effect of those two factors as plasticisers, is similar in all stress forms that induce creep. Clearly also, creep in shear must be governed by the same anatomical features as determine the responses in compression and tension. Further reference will be made to responses to shear in the discussion of wood failure modes.

The concept of mechanical conditioning

The notion of 'mechanical conditioning' of structural materials, which is sometimes called 'work' or 'strain hardening', has long been recognised. For example, Eshbach (1952) repeated the following remarks from his 1936 edition – 'to strain a metal beyond its yield point and then allow it to rest results in what is called *strain hardening, i.e.*, an increase in the value of the yield point. The same action occurs in extremely slow straining, so that sometimes an alloy

* Obtained at the Division of Forest Products, CSIRO, in 1955.

after months of creep will become strengthened by the operation to a point where it will no longer yield to the strain'. Leaderman (1944) discussed the phenomenon of mechanical conditioning in respect of the broader field of high polymers. He too identified it as a change in properties after application of stresses. Also, he noted that many high polymers exhibit an irrecoverable deformation when subjected to a stress for the first time, but later stress cycles do not increase the irrecoverable component, as long as they do not exceed the initial 'conditioning stress'.

Strain hardening is known to occur in plant tissues also. For example, its occurrence was demonstrated with isolated sections of the cell wall of *Avena* coleoptile, on which Cleland (1971) made tensile tests. With reference to the effect of pre-extension on creep rates, he noted that 'the creep rate of coleoptiles at a particular stress is markedly reduced, if the walls have been pre-extended at a greater stress'. In respect of the rate for the second extension, he noted that the position of the change to the higher rate was determined by 'the amount of previous extension, not the amount of previous stress'. The rate of creep then is 'roughly proportional to the stress in excess of that needed to extend the section back to the length reached during the pre-extension'.

Similarly for stress responses in wood, Kingston (1962) stated 'repeated tests on the same specimen, loaded to the same stress in bending, have shown a marked decrease in the observed total creep in successive tests. This phenomenon is known as mechanical conditioning'. Also, he suggested that 'time effects on deformation will occur in the non-crystalline region of the cellulose rather than the crystalline region or in the lignin. For similar reasons, it may also occur in the regions rich in hemicelluloses on the borders of the lignin in the primary or outside layers of the fibre walls.'

Generally, all those data and related deductions and postulations are compatible with the discussion herein, on effects of imposed stresses on characteristic features of the fibre wall, and particularly on the interaction of the microfibrillar framework and the matrix material. Thus, when the applied forces have caused modifications to the configurations of the microfibrils, and corresponding viscous flow of the matrix material (as was deduced earlier), there could not be a complete return to the precise anatomical dispositions prior to the imposition of forces.

For an analysis of strain hardening, it may be assumed that the force which produced the creep is removed, and that no special treatment is applied to stimulate creep recovery. The elastic strain recovery would tend to modify somewhat the disposition of the matrix prior to removal of the imposed force, but that would require time. However if a similar force were quickly re-imposed, there should be no significant creep response until the strain reaches the maximum previously developed. Thus the wood fibres would be expected to show the characteristics of strain hardening. Correspondingly, the fibres would exhibit elastic responses (without creep) up to a considerably higher stress level than in the initial situation. Restraining up to the initial extent would involve only elastic straining of the crystalline microfibrillar framework of the cell wall, and a static reaction of the matrix.

Anatomical data related to creep and mechanical conditioning

Murphey (1963) reported that, when specimens were subjected to tensile loading parallel to the grain, there was an *immediate* increase in crystallinity of the cellulose chains, and apparently that was related to the intensity of the applied force. He considered that the immediate effect then remained constant with sustained loading. Hill (1967) reported on comparable X-ray diffraction evidence of an increase in orientation of holocellulose in pulp fibres subjected to a sustained tensile force, but did not study relationships to time. Based on such data, Murphey (1963) and Schniewind (1968) concluded that 'any structural changes accompanying creep must take place in the unoriented regions of the cell wall'. Mark (1952) suggested that, when subjected to tensile stresses, the effect of the strain on the poorly-oriented cellulose molecular chains in the 'amorphous' zones in the microfibrils may disentangle and cause them to 'gradually become parallel. This leads to an additional crystallization'. Those postulations are discussed below.

Murphey's data have shown that tensile loading increased the range of elastic response of the wood to tensile forces. However, doubt attaches to his conclusions that; (1) the significant changes relate specifically to the unoriented regions of the wall; and (2) that the cause of the changed response is that 'additional cellulose molecules are brought into alignment, thus increasing the length of the crystallite and therefore the degree of crystallinity'. In respect of the first point, it should be clear from earlier discussion herein, that effects could not be restricted to unoriented regions of the microfibrils: a general interaction between the crystalline cellulose microfibrils and the matrix and encrusting materials must be involved in any changes within the cell wall.

Murphey's second conclusion is related to his statement that 'the cellulose chains are recognised to have areas of crystallites and disoriented regions', which clearly is a reference to the 'fringe micellar theory'. However, it seems certain that if there were such regions of non-oriented and widely separated chain molecules of cellulose, which the fringe micellar theory postulates are largely discontinuous through that region, they would be encrusted with lignin and all spaces between them would be filled with the matrix material. It cannot be conceived that individual molecular assemblies, in such a physically disordered group of chain molecules, can have sufficient tension imposed on their '*loose*' portions or ends, so that their random lengths and dispositions could be re-organised into precisely parallel and uniform, closely-spaced molecular arrays. Even if tensions could be applied, the presence of the encrusting material, plus the bulk separation effect of matrix between individuals, would prevent such a random arrangement of dispersed molecular chains being drawn together into the very precisely-spaced and perfectly ordered arrangement which is characteristic of crystallisation. Furthermore Murphey's implication – that such extensive reorganisation of the molecules could be stimulated by the lower stress intensities imposed during those tests – seems especially improbable.

It seems more probable that Murphey's data showed that the axial tensile stresses increased the intensity of X-ray diffraction, simply because they induced an increased uniformity of direction of the microfibrils (with their *pre-existing, basically high crystallinity*). That would be compatible with effects of tension as noted by Frei and Preston (1961) and by Balashov *et al.* (1957). With microfibrils in the S_2 oriented in a near-axial direction, even the lowest of the imposed tensile strains would cause some narrowing of the openings between the microfibrils in lenticular trellis configurations, and consequently would have such an effect.

Thus the intensification of X-ray diffraction would be an inevitable result of some straightening of microfibrils, as a natural physical response to the application of an axial tensile force. That would occur in the absence of the postulated increased crystallisation, for which there is no evidence or convincing argument, particularly for its occurrence at low stress intensities. The improved uniformity of alignment of the microfibrils would necessarily involve a corresponding change in the outline form and location of the associated matrix material. Consequently, if the specimen were relieved of the imposed force, and reloaded soon afterwards, it would exhibit an enhanced range of elastic behaviour, as noted by Murphey. Thus the concept of re-alignment of otherwise unaltered microfibrils (in respect of crystallinity) would account also for the enhanced elastic behaviour, as a consequence of strain hardening effects.

It is notable also that Hill (1967) reported an increase in orientation of holocellulose fibres under creep load, and that the observations by Yamada *et al.* (1966) appear to be compatible. Some estimates of the effects of stress and creep, on the mean microfibril angle in the S_2 layer of the fibre wall, have been made on the diffractograms* published by Murphey (1963). The estimates indicate that,

---

* These diffractograms are very far from ideal for the assessment of microfibril angle. Particularly, that is because the wood specimens were cut in an axial plane located half-way between tangential and radial; the extremely asymmetric peak forms were a consequence. The very small scale of the 'no-load' diffractograms also seriously prejudiced accurate assessments, as indicated by Boyd (1977b). Nevertheless, the estimates based on the method developed by Boyd (1977b) yielded values which appear qualitatively indicative, of an improvement in microfibril alignments though not reliable in respect of precision.

compared to the 'no-load' situations, Murphey's '100 per cent' load had the immediate effect of decreasing the mean microfibril angle in S_2 by $c.0.5°$. Then the creep effect of that load over 24 hours appeared to cause a further decrease of $c.1.2°$. After removal of the load, the residual increase in mean orientation (relative to axial) appeared to be $c.1.5°$.

Regrettably, the reproduction of the diffractogram for 'after load removal' is very poor, and it results in the estimation of mean microfibril angle being of doubtful precision and reliability. Nevertheless a comparison of that estimate, with the one from the 'no load' diffractogram, indicates there was a slight residual creep effect immediately after load removal. The indicated amount of creep under full load suggests that the residual, after unloading, should be a greater proportion of the initial creep than is shown by the figures above. Overall however, the analysis is compatible with the loading increasing the perfection of orientation of the microfibrils. Thus it would account for the increased intensity of X-ray diffraction, without recourse to the very questionable postulation of increased crystallisation.

That Murphey's data indicate an extension of the range of elastic response, on subsequent load application, tends to confirm the vague indication of a small permanent modification of the configuration of the microfibrils, towards parallelism in a direction slightly nearer axial. With tensile tests on individual fibres, Hill (1967) reported clear evidence of these effects. Under the influence of creep, there was a substantial reduction in the microfibril angle, in the S_2 layer. On removal of the tensile force, the angle increased (recovered) by about half the change caused by the earlier loading (and creep).

Hill's data on tensile loading showed also that enhanced values of physical properties occurred in association with both the effective increase (deduced) in parallel arrangement of microfibrils, and the reduction of microfibril angle as a consequence of creep. Thus the residual mechanical conditioning effect led to large and significant gains in modulus of elasticity and rigidity of the fibres. That effect increased with the duration of the loading, and with the increase in applied tensile force, by proportional amounts up to $c.70$ per cent. That consequence is compatible with the previously discussed nature of interactions and relative rearrangements of microfibrils and matrix materials, under the influence of tensile forces.

Effect of load duration on ultimate strength of wood

It has been established (Wood, 1951) that the ultimate strength of wood in bending is reduced as the duration of loading increases. For example, for constant environmental conditions, Wood's data indicate that the ultimate stress at failure in impact loading is about twice that for a load applied over a period of 10 years. Correspondingly, it is known that the creep or deflection up to the point of failure increases with the load duration. These facts can be related to the effect of load duration on the interaction between the microfibrils and matrix materials in the fibre wall.

In respect of axial compression of wood under the influence of a particular applied force, it has been shown herein that, with increasing duration of loading, there would be increased flow of matrix material within the walls of the fibres. Associated with that, there would be increased space to accommodate the matrix between microfibrils. Thus instead of the initial firm, continuous packing of the matrix around microfibrils, there would be reduced matrix-to-microfibril contact and interaction. Therefore, since the microfibrils are loaded axially, they would receive reduced support against lateral deflection. Hence their stiffness as columns would be reduced, their lateral deflection would increase, and there would be a resultant, highly significant increase in the bending stress and total stress on them. Inevitably that must increase the probability of microfibril and fibre failure, as the duration of loading increases. Additionally, the force required to cause failure in compression would be less, as the duration of force application increased.

In respect of tensile axial forces, the reactions of the fibres are quite different. As the period of load application increases, divergences of microfibrils from their mean direction would decrease. The bending stresses on the microfibrils would decrease correspondingly, and thus cause a reduction in the total stress on the microfibrils. This suggests that the microfibrils and the fibres should be able to sustain higher applied forces. Hill's (1967) data demonstrated that there was such an increase in tensile strength, by up to 30 per cent, depending on the increase in intensity of stress application and its duration.

In the unseasoned condition, water fills the pores in the matrix and elsewhere between microfibrils. Hence, the material enclosed by the microfibrils tends to act as an incompressible body. The pressure exerted on the matrix, as it is compressed in its shuttle-like shape between the tensioned microfibrils, would tend to force some water out of the pores; but it is suggested that the extent of that action is likely to be small. Correspondingly, the increase in strength in tension would be small in comparison to the decrease of strength which occurs under equal compressive forces.

If the wood were subjected to desorption either before or during application of the tension, the loss of water would have the effect of compacting the matrix material and the microfibrils would be drawn into improved alignment. However, the increased viscosity of the matrix would resist additional compaction under loading. Hence, it appears that the effect on tensile strength properties, as the duration of loading increases, is likely to be an increase, as with the un-seasoned material. In fact, Hill's (1967) data show that material stressed during seasoning develops increases in modulus of elasticity and rigidity that attain the same levels as those for unseasoned fibres.

The strength of a bending specimen, which is subjected to long-term loading, must be a composite of the effects of compressive and tensile stresses, together with a small influence of shear stresses. The overall effect on strength must be dominated by the effect in compression. Hence, strength must steadily de-crease in parallel with the increase in deflection or creep, as was discussed before, when dealing with the effect of bending stresses. This deduction, which is based on interactions of matrix and microfibril components of the cell wall, is compatible with published data such as those of Wood (1951).

Effect of temperature level on creep

Stamm (1964) made a general reference to the fact that a temperature increase had a plasticising effect on the reaction of the cell wall. Davidson (1962) dem-onstrated that, for two softwood and one hardwood species at moisture con-tents in the range from 18% to 20%, the rate of creep increased with tempera-ture level (held constant). Comparable rate increases occurred with each 10°C step from 20° to 50°C, but there was a much larger rate increase with the step from 50° to 60°C. It was suggested that the latter might be explained if 'some of the hydrogen bonds in the wood cellulose were broken at those higher tem-peratures'.

In bending tests on Japanese cedar soaked in water, Kitahara and Yukawa (1964) showed a similar increasing rate of creep between successive steady tem-peratures in the range from 20° to 50°C (50°C was the highest temperature in-vestigated). With the same species and moisture conditions, Arima (1972) made tests in compression perpendicular to the grain, at temperature levels from 20° to 75°C. In that case there were similar initial rate changes with temperature steps, but a substantial increase in creep rate at the higher temperatures; most of the latter seemed to be stimulated at *c*.60°C. Similar creep responses of hoop pine to temperature were demonstrated by Kingston and Budgen (1972), for the range from 20° to 50°C. Comparable effects on the torsional stress relaxa-tion of wet Hinoki wood were shown by Urakami and Nakato (1966). All of those data are compatible with the previously described interaction of matrix and microfibrils within the fibre wall, as associated with a temperature increase leading to reduced viscosity of the matrix.

Elastic recovery of visco-elastic creep

There are many experimental data showing that, if the forces causing creep deformation are removed, there will be an immediate recovery of the elastic component of strain. They show also that this is followed by recovery at a reducing rate, of some of the creep strain. No satisfactory explanation for that recovery of creep strain has been offered. At best it has been described as a 'memory' response, such that the deformed material endeavours to return to its original form. In fact, the creep recovery reaction is readily explained in terms of natural physical interactions between microfibrils and matrix materials in the cell wall.

When the fibres are stressed in either tension or compression, by forces imposed parallel to the fibre axis, the force transferred to the microfibrils must bend them and compress the enclosed matrix material transversally (as explained before). Thus the microfibrils and matrix materials must absorb strain energy, and approach compatibility of form and a balance of forces and reactions in a new configuration. At the same time however, there would be an associated small loss of energy due to microfibril rearrangements involving the breaking of chemical bonds, and due to frictional losses during the forced flow of the matrix. That energy would be irrecoverable. The elastic forces in the crystalline microfibrils would be counterbalanced by the transverse pressure reactions of the matrix materials.

Subsequent removal of the applied forces would allow recovery of elastic strains and the resilience of the microfibrils would cause a return towards their initial configurations. With relief of their pressure on the matrix materials, the strain energy previously stored as a consequence of the compression would be released. Thus the matrix also would tend to return to its original volume, shape and location. In this process, the interaction between those two critical structural factors would tend to maximise the amount of form recovery possible in a given set of environmental conditions.

With other forms of force application, such as bending, compression perpendicular to the grain and shear, comparable interactions would occur between microfibrils and matrix during loading and creep development. Also, on removal of applied forces, similar reactions would stimulate some recovery of the initial form and dimensions. For any particular environmental conditions, creep recovery must be dependent almost completely on the extent to which conversion of strain energy stored in the microfibrils is able to force viscous flow of the matrix towards its initial position. Given sufficient time, all that strain energy would be released and directed towards achieving that flow. If the temperature were raised, that would facilitate flow of the matrix and increase the rate of creep recovery. Moisture cycling can have a similar effect, as will be shown.

Mechano-sorptive creep in wood

Over the last two decades, there have been many intensive experimental studies of this creep phenomenon, which results from moisture changes in wood when it is subjected to an applied force. Such moisture changes may occur as a consequence of normal daily and seasonal variations in the environment. The effects of creep so induced are recognised as of very substantial practical significance in structural situations. Particularly, that is because the deformations involved with extensive and repeated cyclical moisture changes may be up to five times greater than those arising from visco-elastic creep at the same imposed stress level. Consequently mechano-sorptive creep can cause aesthetically unacceptable deflection of beams and columns, and may seriously reduce strength and service life of the timber.

Despite the importance of mechano-sorptive creep, Grossman (1976, 1978) noted that there was no satisfactory explanation of the basic cause of the many different aspects of this phenomenon. Nevertheless, the various manifestations can be explained in terms of the simple physical interactions of microfibrils and matrix materials, which occur during moisture changes associated with stress application. That will now be illustrated. In the process, reference will be made to Grossman's (1976) listing of the most important mechano-sorptive observations, against which a model explanation should be checked for compatibility and adequacy.

Creep dependence on extent but not duration of moisture change

Armstrong and Kingston (1962) and Schniewind (1966) produced many data which showed that mechano-sorptive creep depended on the extent of moisture change, and that it was little affected by the rate of change. Again, that can be explained in terms of the interaction of microfibril and matrix responses.

Consider schematic arrangements of microfibrils, such as illustrated in Figs. 3, 4 and 5, and assume that in the saturated condition of the matrix material (water-swollen condition) it fills all of the spaces enclosed between microfibril to microfibril bonds. When moisture is withdrawn from the cell wall during desorption, the matrix will shrink to a smaller volume, as indicated in Fig. 5. If an external force is applied to the wood during (or after) the moisture reduction, there will be a change in the volume enclosed between the microfibrils, as discussed before. If at any time during this force application there were any difference between the volume occupied by the matrix and that enclosed by the microfibrils, there would be a corresponding reduction in the support which the matrix could provide against the lateral bending tendency of the microfibrils.

The microfibrils are slender, flexible and slightly curved between positions bonded to other microfibrils, but their rigidity would be enhanced greatly if the matrix were in a position to provide lateral support equivalent to only a very small percentage of the axial force on the microfibrils. Experience has confirmed the validity of that statement for structural units acting either as columns or beams (Pearson *et al.*, 1958). In the absence of such lateral support, the externally-applied force would tend to induce severe deformation of the microfibrils, fibres, and wood tissue as a whole. Such deflection must be much greater than with visco-elastic creep. That is because with no reduction in moisture content, the matrix would fill the space between microfibrils, and thus deformations of the microfibrils would be fully opposed by the rigidity of the matrix and its resistance to flow.

Following loss of water from the cell wall, and associated shrinkage of the matrix, the microfibrils would deflect to such an extent that they would contact and apply a local pressure to the matrix material. That would stimulate flow of the matrix to conform to the deflection shape of the microfibrils. Such flow from localised pressure points would allow further deformations of microfibrils, until there was an improved compatibility of form of the two elements, and normal equality of the force applied through the microfibrils and the matrix reaction. Therefore, according to whether the externally-applied force was compressive or tensile, there would be a corresponding imposed flow of matrix material to achieve stabilised forms of the microfibrils, such as in Fig. 4b and 4c.

Thus, during a reduction in moisture associated with the application of an external force, there would be two components of resulting deformations. The first would be a relatively large deformation facilitated by the shrinkage of the matrix; that would involve some consequent loss of restraint on lateral bending of the microfibrils. The second would be an interaction effect which is of the general nature of that involved in visco-elastic creep. The latter effect must always be present during mechano-sorptive creep. However, except during initial loading in the dry condition, its contribution to the total deformation would be small compared to that of several large changes in moisture content below saturation level in the wall.

With reference to Fig. 5, in an unstressed situation the extent of separation between microfibrils and matrix must be dependent solely on shrinkage of the matrix, and hence on the *extent* of the moisture change involved. The period over which that change occurs is irrelevant. Application of an external force, during the moisture change, does not influence the total space vacated by the matrix as a result of its shrinkage. As a consequence of shrinkage of the matrix, initially the imposed force simply causes distortion of the microfibrillar structure, and consequently it facilitates a renewed contact between microfibrils and matrix, or more generally it tends to maintain continuity of contact.

The period of re-adjustment, arising directly from the moisture change, depends to some extent on the magnitude of the externally-applied force. The rate of moisture change (in a cyclical sequence) has a small associated effect. It determines the period available for the externally-applied force to cause viscous flow of the matrix, and thus induce balanced interaction between the forces in the microfibrils and the induced responses by the matrix. Overall however, the dominant factor which determines the creep deformation is the total moisture change, and not its rate or the period involved. Generally, the mechano-sorptive creep response will occur as rapidly as the moisture change.

Initial deformation increase with either a rise or fall in moisture

Whatever the loading mode (compression, bending, tension or shear), there will be an increase in creep with the initial change in moisture content of the cell wall, whether that change be an increase or a reduction. That has been demonstrated by data such as those of Armstrong and Kingston (1962) and Schniewind (1967). This response also has not been explained previously. Nevertheless, it is explicable in terms of the corresponding imposed interactions of the microfibrils and matrix material.

Initial reduction in moisture content. − As a consequence of a reduction in moisture content, the matrix would shrink as shown schematically in Fig. 5 and 7a. Then if the externally-imposed force were compressive, the matrix would be impelled to flow in the direction of that force, as illustrated in Fig. 7b. At the same time, there would be an additional bending tendency and an increased axial deformation imposed on the enclosing microfibrils. Although they may not be forced into rigid contact with the matrix at its upper boundary, they would quickly reach a stable configuration. That would occur because of the interaction of adjacent microfibrils and included matrix material. The random arrangement of bonded positions, along contacting microfibrils, would ensure that external support was given at the top of lens-like units such as the central one in Fig. 7a & b.

If there were an externally-imposed tension associated with an initial shrink-

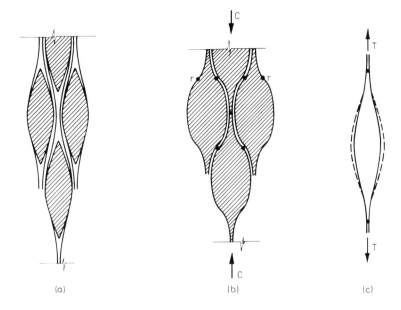

Fig. 7. Combined effects of moisture change and applied force on microfibril configurations. (a) Effect of moisture loss, without force application. (b) Effect of moisture loss, followed by compressive force application, and subsequent regain of moisture. (c) Effect of moisture loss, followed by tensile force application, and subsequent adsorption. For the original non-stressed condition, the composite profile of the microfibrils is indicated by a broken line.

age of the matrix as in Fig. 7a, that would draw the microfibrils into a more parallel arrangement. Hence, they would 'squeeze-in' on the matrix material as their deformation increased, as illustrated in Fig. 7c. Thus tensile creep deformation (extension) would occur. On the concave and convex sides respectively of a beam, those compressive and tensile effects would be involved during bending. However, the influence of compression on the concave side would greatly out-weigh that of tension on the other side. Hence, the overall effect would be a deformation increase which approached that in compression. Other loading modes would lead to comparable modulation of creep effects.

Initial increase in moisture content. – If during a creep test, the initial moisture content of the cell wall were below saturation level, the configuration of the microfibrils and the location of the contracted matrix material would be related to the previous loading of the wood. In general, it would be related to material sawn from a green log after felling the tree. Therefore, on the *im-*

mediately prior occasion the wood would not have been subjected to any substantial stresses, except those which may be associated with drying. Assuming an absence of distortion of fibre shapes due to collapse during drying, a subsequent sorption, up to fibre saturation level, in the absence of applied forces, would restore the matrix to its volume and form prior to desorption. However, during desorption of normal wood there is virtually no shrinkage in the axial direction of the fibres; correspondingly , there would not be axial extension with sorption of moisture causing a re-swelling of the matrix.

Hence, if that first sorption phase occurred at the same time or after applying a force in compression parallel to the mean orientation of the microfibrils, they would tend to deflect laterally and reduce the axial distance between bonded positions, with the effect as indicated from Fig. 4a to Fig. 4b. That would provide more space for the matrix than was available in the original green condition. Hence, with swelling of the matrix to its volume prior to the original desorption before loading, it could not resist significant axial deformation of the system. Accordingly, deflection of the microfibrils would occur until constrained by the swollen bulk of the matrix, or by lateral space restraints imposed by adjacent contacting microfibril and matrix systems, *i.e.*, compressive creep would occur.

Similarly, if after initial desorption without force application, axial tension were applied at the same time or before sorption occurred, the microfibril system would tend to extend (deform) as between Fig. 4a and Fig. 4c. Indeed some tendency to approach that configuration would have occurred as a consequence of surface tension forces operating during the initial desorption (Stamm, 1964). That would be augmented significantly by strain due to the externally applied tension. During sorption the swelling of the matrix could not fully offset the change in dimensions as from Fig. 4a to Fig. 4c. That is because the tensile force 'T' would cause an immediate axial extension and transverse contraction. That would oppose swelling of the matrix, leave less space for moisture to enter the system and thus restrict the potential of the matrix to swell.

Correspondingly, both the effect of compression on the concave side and tension on the convex side of a beam under load would contribute to creep when the first moisture change was sorption. In all subsequent moisture cycling however, sorption would lead to some reduction in creep and desorption to an increase as has been observed, and as was explained earlier herein.

Alternating increase and recovery of deformations with moisture cycling under stress

Extensive data show that, except the increase in deformation when wood is initially subjected to a rise in moisture content under an applied force, each reduction in moisture content leads to an increase in deformation, and each increase in moisture leads to some reduction in deformation (*e.g.* Armstrong and Christensen, 1961; Armstrong and Kingston, 1962; Hearmon and Paton, 1964; Bethe, 1969). The general reason for an applied force causing a large increase in deformation, during a moisture reduction phase, has already been outlined in the previous section with particular reference to the first in a cycle of moisture changes. Hence, consideration will now be given to reactions during the typical moisture increase in a cycle of reductions and increases.

Typical creep recovery during moisture sorption. – Associated with a reduction in the moisture content of the cell wall, the microfibrils deflect in a manner such that they tend to establish a firm contact over the contracted boundaries of the lens-shaped matrix between them. With the subsequent sorption of water, the matrix must swell. The swelling force must push the microfibrils back towards their configuration prior to the earlier reduction in moisture content. Such a recovery tendency is inevitable because the major part of the creep developed during desorption is a consequence of the loss in volume of the matrix, and the resulting absence of constraint on the deflection of the microfibrils.

However, there cannot be a complete reversal of the previous deformation phase. Partly that is because of the energy dispersed in friction and severence of bonds. Perhaps more importantly, it is because of a very significant difference between the resultant of active forces during the desorption phase, and that during sorption.

For example, if we consider a compressive force effect during a desorption phase to be represented as in Fig. 4b. This is a consequence of shrinkage of the matrix, deformation of the microfibrils, and associated shortening of the axial distance between successive bonded positions along their length. That involves a change in the energy balance, due to a movement in the direction of the force, of its position of application. Effectively the position of the force and reaction coincide with successive axial positions of bonds between adjacent microfibrils as in Fig. 4b. The change in position (movement) of the force is equivalent to conversion of potential energy to induce the new configuration.

On the other hand, complete restoration of the previous microfibril configuration, with subsequent swelling of the matrix, would be opposed by the compressive force and involve a requirement to change its point of application back to the original position. The direct effect of the compressive force would be to restrict the extent to which water could be absorbed by the matrix. Correspondingly, it would restrict the potential swelling capacity of the matrix. Additionally, part of the swelling force actually developed by the matrix would be offset by friction forces associated with viscous flow of the matrix during swelling. Hence the net swelling force could not compel a return of the point of application of the compressive force to its position prior to desorption. Accordingly, although sorption would lead to some reduction in strain (creep recovery), the residual creep would be greater than that before any previous desorption and sorption cycle.

Progressive reduction in creep increments during moisture cycling. — Generally the physical factors involved lead to each successive creep increment on desorption, and each successive recovery with sorption being less than the respective previous one. Perhaps that is most obvious from consideration of tensile creep. As a basis of analysis, assume that the matrix material fills or approximately fills all the lens-shaped openings between microfibrils in bonded arrangements, as in Figs. 3 and 4, but that it is not necessarily of uniform density within them.

Following desorption and shrinkage of the matrix as in Fig. 5, the tension would tend to straighten and draw-out the microfibrils between bonded positions. That would apply transverse pressure to the enclosed matrix. The resulting densification, together with the loss of moisture, would increase viscosity of the matrix and so restrain flow to fill the available space. Subsequent sorption would facilitate such flow. However, it has been discussed how full recovery of the previous deformation increment during desorption cannot occur under continued loading, and therefore the space available to the swollen matrix would be less than before. Overall then, there must be an increase in mean density of the matrix and less space for water. Hence in each successive desorption phase, there would be less water loss and less shrinkage and less forced deformation (creep). Correspondingly there would be progressively less recovery in successive sorption phases.

With the compressive mode of creep development, the restraints are somewhat different. As a consequence of desorption, the effect would be to progressively reduce the axial distance between bonded positions on the microfibrils, as illustrated by a comparison between Figs. 4a and 4b. Simultaneously, there tends to be a lateral spread of the space between the microfibrils, and that would increase with additional deformation associated with each desorption phase. However, the similar effects in laterally-adjacent lenticular trellis configurations would lead to increasing constraint. The microfibril arrangements at the boundaries of the fibre wall are perhaps even more significant in respect of constituting constraint on broadening of the lenticular space.

The orientation of microfibrils in the outer layer (S_1) of the fibre wall, and also in the innermost layer (S_3) is close to transverse. Hence the helical turns of the microfibrils tend to act as hoops. The axial compressive force tends to induce a transverse spread between microfibrils in the main structural layer (S_2), and thus tends to thicken it. As a consequence, radial pressure would be applied to the S_1 and S_3 layers. However, displacement towards the lumen side is resisted not only by the arched form of the S_3 layer in the transverse direction, but also by the fact that movement in that direction involves a reduction in the circumference of the inner lamellae of microfibrils. Hence microfibrils and matrix would tend to be squeezed to a reducing overall lateral spread. That must oppose the opposite tendency arising from the imposed axial force. Perhaps even more positive resistance would be offered to expansion radially outwards, as that would stress the transverse microfibrils in S_1 in tension, *i.e.,* in their mode of greatest strength and rigidity.

That the tensile strength of S_1 effectively opposes transverse expansion tendencies is well illustrated by micrographs such as some reproduced by Roelofsen (1959). After an extreme, chemically-induced swelling of the fibre wall in wood, *Pandanus* and cotton, the microfibrils show ballooning of the S_2 layer (or its equivalent in microfibril arrangement) between intact sections of the partially destroyed outer layer of transverse microfibrils. Obviously an intact S_1 wall structure must place very substantial restraint on transverse expansion arising from effects of an axial compressive force during a desorption phase. Hence creep so induced would be in reducing increments with successive desorption phases.

These deductions of a tendency for reducing creep increments, resulting from moisture changes associated with actual imposed tensile, compressive (and bending) forces, are compatible with actual creep measurements such as those by Armstrong and Christensen (1961), Armstrong and Kingston (1962), Hearmon and Paton (1964), and Bethe (1969).

Relative effects of loading modes on extent of creep

Extent of creep under tensile forces. – In the immediately-prior subsection, it has been pointed out that an imposed axial tensile force reduces the volume available to accommodate the matrix material between microfibrils. Hence with successive moisture cycles, the increasing densification of the matrix, under the influence of tensile forces, must result in a continuously reducing space for accommodation of water, and therefore cause increasing resistance to further straightening of the microfibrils and further creep. Thus from the first creep associated with moisture change, successive creep increments in tension must be of continuously reducing magnitude.

Extent of creep imposed by axial compression. – Because of the curvature of the microfibrils, as represented in Fig. 4a, an axial compressive force induces an increased outwards deflection of the microfibrils, as in Fig. 4b. Hence, progressively the openings between adjacent microfibrils tend to approach spherical forms, and thus increase their enclosed volumes. Accordingly, there is not a direct limitation of reducing space for the matrix, such as causes restraint on creep in tension. However, as indicated before, an overall effective constraint on transverse expansion develops at the boundaries of the cell wall.

On the other hand, due to the relatively large thickness of the S_2 layer, there is a possibility of considerable accommodation of transverse expansion by adjacent lenticular forms. As discussed earlier and illustrated in Fig. 7a, the random arrangement of adjacent forms limits the outwards spread of microfibrils under compression, in the region above the enclosed matrix. Therefore there would be a tendency to force the adjoining microfibrils towards the matrix in the unoccupied space above it, as indicated in Fig. 7b. Thus some of the transverse expansive tendency in adjacent lenticular forms would lead to curvature reversals, and thus to increasingly unstable shapes of microfibrils as indicated at positions 'r' in Fig. 7b. In turn, that would predispose them to local buckling or crinkling at such positions.

Overall it is apparent that, with imposed axial compression, much greater mechano-sorptive creep would occur in each desorption phase, than is the case with imposed tension. In the long run however, increasing constraint would reduce the extent of creep and recovery increments in compression, during successive desorption and sorption phases. As indicated earlier, there will be similar effects, of intermediate magnitudes, for creep associated with other loading modes. All of the foregoing deductions on creep, arising from interactions between imposed force and mode, microfibril responses, and effects of moisture changes within the matrix material, are compatible with the various measurements of mechano-sorptive creep, such as those of Armstrong and Kingston (1960, 1962), Bethe (1969), and Erickson *et al.* (1972).

Creep recovery after force removal and moisture cycling

It has been noted that, of creep deformation which remains after removal of the primary activating force, a large part is at least temporally 'frozen in' (Christensen, 1962). However, much of that is recoverable if the wood is taken through another moisture cycle (Armstrong and Christensen, 1961; Christensen, 1962). That recovery appears to be a consequence of removal of constraints resulting from the force-imposed-relocation of the matrix material.

When the wood is in a moist condition at the end of the loading period, a desorption phase would shrink the matrix. That would remove constraint which the swollen form and position of the matrix imposed on elastic recovery of the original shape and position of the microfibrils. However, such a single phase treatment is unlikely to enable the maximum practicable recovery in a reasonable period, of large deformations resulting from mechano-sorptive creep. In a subsequent sorption phase, the matrix would tend to swell back to the same form as prior to desorption. Achievement of that form would require that the swelling pressure of the matrix force the microfibrils back to their previous configuration, which was a consequence of the earlier loading. The rigidity of the microfibrils would resist such deformation; but their reaction could develop only as a consequence of some deformation being forced on them.

While swelling of the matrix would thus be restrained in the direction transverse to the microfibrils, its flow in the axial direction of the microfibrils would be unimpeded. Consequently the lateral reaction of the microfibrils would cause such a flow of the matrix. As the swelling pressure of the matrix develops gradually, microfibril reactions would stimulate some flow of the matrix from the initiation of contact between the two elements. However, on completion of sorption the microfibrils would be deflected somewhat from their location prior to the original load application. With sorption during a repetition of moisture cycling, substantially less swelling pressure would be exerted by the matrix. Thus there would be an improved final return towards configurations of microfibrils and matrix prior to loading, *i.e.*, there would be additional creep recovery.

If the wood were in a dried condition when the force causing creep was removed, a sorption phase would facilitate faster flow of the matrix, as it is stimulated by elastic strain from the deflected microfibrils. However, maximum recovery may be spread over an extended period, because of the slow viscous flow of the matrix. The addition of a desorption and sorption cycle would enable faster recovery. Several cycles may be necessary to achieve the maximum total recovery. The recovery mechanism thus described is compatible with the data reported by Grossman (1953), Kubler (1973) and Tokumoto (1973).

As will be discussed in relation to the failure process in fibres subjected to compressive forces, the development of large deformations is known to involve formation of microscopic crinkles in the microfibrils and in fibre walls. With increasing compressive deformation, it has been demonstrated (Dinwoodie, 1968) that the crinkles increase proportionally in number and severity. Accordingly much of the strain energy, which was stored in the microfibrils during their elastic deflection, would be dissipated as a consequence of breaking of bonds and elastic straining to allow rearrangement of the molecules of cellulose, as would be necessary to attain the crinkled form of the microfibrils.

After removal of the compressive force, swelling of the matrix to recover creep strain during a sorption phase would tend to straighten out such crinkles; but inevitably much energy would be dissipated in the process. Because of the factors mentioned above, several sorption cycles would be necessary to achieve maximum recovery. Even so, that recovery could never be complete. Furthermore, where the amount of creep is large, as with moisture cycling while subjected to compressive or bending forces, a corresponding severity of crinkling is to be expected (see below). Consequently the proportional recovery would be less, as compared to that after visco-elastic creep at moderate force intensities.

Other observations related to creep

Several additional factors have been involved in studies of creep responses. These include swelling of the matrix by substances other than water, the 'bond slippage' hypothesis, temperature level during testing, wood species, and cell wall collapse associated with desorption. Brief references will be made to their significance.

Swelling agents other than water

Apparently typical mechano-sorptive creep responses in wood were observed when some organic solvents were used to replace water, but not with others (Gardner et al., 1969; Bach, 1974). Clearly their data were ineffective in identifying the basic factors which determine creep responses; but at least they had a bearing on the validity of some hypotheses proposed previously to explain this phenomenon. For example, although hydrogen bonding had been proposed as a significant mechanism controlling creep responses, data relating to the effect of dioxan did not support that. Similarly, the differences in relative creep effects of methanol, ethanol and propanol were 'surprisingly large' in comparison to the small variations in their dielectric constants. That suggested that other factors were significant, and particularly 'affinities towards lignin' (or matrix substances as designated herein).

The bond slippage hypothesis

Armstrong and Christensen (1961) and Christensen (1962) reported that, when a 'specimen in an evacuable apparatus was loaded in the complctely dry state, very little creep occurred within a short time of loading'. If then 'water vapour at saturation pressure was admitted, the deflection increased rapidly to almost twice the initial value'. That effect may be compared to that of increased temperature levels as discussed earlier. Either singly or in combination, both moisture and temperature would affect the plasticity of the matrix, and thus affect its creep reaction to stresses imposed through the microfibrils, as has been discussed earlier herein.

Grossman (1976) mentioned some authors who had favoured the hypothesis that during the breaking and remaking of bonds associated with changes in moisture content within the fibre wall, an unbalanced stress system caused slippage between microfibrils and matrix, *i.e.* caused creep. Additionally it was proposed that the amount of movement of moisture through wood controlled the extent of creep response. However, Grossman also remarked that Armstrong's (1972) data and conclusions showed that water movement *per se* was neither a cause nor a determinant of the amount of creep.

Comparisons between species

The combined data of many experimenters showed that, including both angiosperms and gymnosperms, qualitatively there was no difference in the type of creep response for all of them. That appears to be fully compatible with the interaction between microfibrils and matrix being the prime determinant of creep reactions. Nevertheless, support for this identification of the basic structural interaction may be slightly clouded. That is because some species tend to develop cell wall collapse and exceptionally large creep during loss of moisture. The significance of the latter interaction is discussed below.

Creep associated with cell wall collapse

In wood prior to seasoning, the volume of water above that causing saturation of the cell walls, must be located in the lumen of the cells. That 'free water' in the lumen cannot affect the extent of bulking of the cell wall, and correspondingly its removal would not generally cause volume changes in the cell wall, as when moisture is withdrawn from the matrix. Hence, if wood is in a stressed state when 'free' water only is being removed, there should be only visco-elastic deformation, unless another critical factor becomes operative. However,

Armstrong and Kingston (1962) noted that some species showed exceptionally large creep deformation in those circumstances. Work by Erickson and Sauer (1969), Erickson *et al.* (1972) and Chen (1974) showed similar effects, which were demonstrated to be associated with collapse of the cell wall.

Heartwood formation involves deposition of extractive and incrusting materials in the cell lumen and within the cell wall. At that time, the pit membranes may become clogged with those materials. In the latter event, significantly reduced permeability causes drying problems (Resch and Ecklund, 1964; Stamm, 1964), and may lead to cell wall collapse. Apparently the collapse phenomenon results from an elastic instability of the cell wall, induced by development of a large suction pressure during withdrawal of free water from the lumen. Erickson (1968) supported Tieman's (1944) conclusion, and claimed that data analyses confirmed that such a force development in the lumen was responsible for collapse.

Freezing timber before drying it has been practiced as a means of avoiding the severe shrinkage and distortion of boards that results from collapse (Erickson, 1968). He recorded very high extractives content among redwood boards, and a high correlation between that and shrinkage and collapse. On the other hand, he noted that prefreezing speeded up drying of the heartwood, 'eliminated those correlations', and led to greatly reduced shrinkage and collapse. He recorded also that, in respect of water being withdrawn from the cell wall, prefreezing made no difference to associated shrinkage.

Erickson and Sauer (1969) tested redwood heartwood beams drying from the green state in an atmospheric temperature of 65°C. They showed that the effect of prefreezing, in reducing creep, was similar to its effect in reducing shrinkage and collapse. Erickson *et al.* (1972) noted similar responses at 65°C and 38°C. They noted also that when extractives were leached from collapsed redwood, a subsequent reconditioning treatment produced a reduced recovery of specimen cross-sections. Apparently that result is related to elimination of the bulking effect on the cell wall, which Nearn (1955) related to the extractives. At the same time it indicates how other bulking factors, such as the matrix materials and moisture, affect the dimensions of fibres in situations where moisture content is changed, as during mechano-sorptive creep.

In another study of collapse and creep in redwood, Chen (1974) showed that ultra-structural differences existed between untreated, and prefrozen, unseasoned redwood specimens, which subsequently were dried at 65°C. He noted that as a result of the freezing, 'a large percentage of the (pit) tori contained a crack, and the tori were not aspirated'. Therefore, despite the presence of large amounts of extractives, water could be withdrawn directly and easily from the cell lumen. Hence, large suction pressures were not developed during drying, and collapse did not occur.

It was suggested that the greatly reduced creep was attributable to elimination of the need to pass the water from the lumen through the cell wall during drying. It was envisaged that such bypassing of the water reduced 'slippage' within the cell wall structure to a corresponding extent. However, Armstrong (1972) demonstrated that 'slippage' due to bond severence did not occur as a consequence of passing water through the wall, when it was subjected to imposed forces; thus when the moisture content of fibre walls was maintained at saturation level, continuous flow of water through them did not change the nature of the creep response from that of typical visco-elastic action to mechano-sorptive action.

If pit membranes are clogged with extractives, or pit tori are aspirated, then in the absence of pre-freezing before drying the wood, the free water in the lumen must be withdrawn through the fibre wall. However, that does not alter the fact that the percentage *moisture change within the cell wall,* during the desorption process, is the same as if that free water did not pass through the wall. Accordingly, the creep due to the passage of water through the initially saturated cell wall can be due only to the final net change in moisture content of that wall as demonstrated for non-collapsing timber by Armstrong and Christensen (1961), Armstrong and Kingston (1962), and Armstrong (1972).

Armstrong and Kingston (1962) demonstrated that, in a collapse-prone species, there is an increment of deformation of wood that is associated with collapse during desorption, while the moisture content of the wall is at saturation level. Erickson and Sauer (1969) and Erickson *et al.* (1972) noted comparable responses with collapse-prone redwood heartwood. Clearly however, that increment above normal responses for visco-elastic creep must be due to collapse deformation of fibres, rather than due to consequences of the typical mechano-sorptive creep reactions. That is apparent from many micrographs in the literature, that show grossly distorted fibre forms, such as may be induced by large, externally-imposed compressive forces in the transverse direction of the fibres, or by wall collapse caused by large internal suction forces.

Extreme deformation during mechano-sorptive creep and the effect on failure

It has been noted widely that, even with quite moderate imposed loading, either repeated moisture cycling over a short period, or very long-term exposure to visco-elastic creep can lead to significant structural damage and to failure of timber structural members (*e.g.* Armstrong and Kingston, 1962; Hearmon and Paton, 1964; Schniewind, 1967; Bethe, 1969; Bolton *et al.,* 1974). Perhaps the best documented evidence on this phenomenon relates to tests in bending and compression. To simplify discussion, the general nature of this effect will be examined with specific references to compression and tension parallel to the grain.

Development of cell wall deformation in compression

In an intensive study designed to identify the most significant characteristics of progressive development of structural damage to the fibre wall, during loading of wood in compression parallel to the fibre axis, Dinwoodie (1968) made some very important observations. After loading both green and dry specimens over a short period, so as to produce a wide range of known stress intensities, his extensive data led to the following conclusions: (1) at stresses less than 25 per cent of the ultimate failure load in compression, there was clear evidence of the development of 'slip planes' across fibre walls; these appear as fine (microscopic) lines of disorientation of the microfibrillar structures; (2) he confirmed evidence of an earlier phase of those local distortions (Kisser and Frenzel, 1950); apparently that took the form of 'slight local thickening in the cell wall formed as a result of small deformations of the fibrils'; these were described as 'stress or thrust lines'; any one of them does not necessarily traverse the thickness of the S_2 layer. It is appropriate to note that the earlier description herein, of the manner in which microfibrils and matrix were expected to react to imposition of compressive forces parallel to their mean orientation is completely compatible with those observations in 'thrust lines'; (3) the 'thrust line develops into the slip plane, many of which develop progressively and horizontally into small and then successively larger creases up to failure, where considerable buckling and delamination within the wall occurs'; (4) 'since the dislocations represent a crinkling of the microfibrils *with resultant shortening of the cell wall,* and there is a progressive increase in their number and size with increasing load ....*these dislocations contribute considerably to plastic deformation'* (creep); (5) 'the recorded increase in number of dislocations in green compared with dried timber is reflected in the higher plasticity of the former'; (6) 'the angle at which the slip plane traverses a cell wall ....is related to the lamellate structure of the cell wall causing differential elasticity in the horizontal and vertical planes'; and (7) finally, the most severe 'dislocations develop preferentially in that part of the tracheid which is in contact with the ray'.

As explained before and illustrated in Fig. 7b, the imposed compressive force induces progressive changes in the shape of adjacent microfibrils, between bonded positions along their length, and also in the shape and location of the enclosed matrix material. Ultimately transverse reactions, from adjacent groups of microfibrils subjected to similar changes, must tend to increase the sharpness of curvature reversals, at positions such as are indicated by 'r' in Fig. 7b. Simultaneously, the application of an axial compressive force on the microfibrils would naturally lead to the transverse swelling effect which was noted by Kisser and Frenzel and by Dinwoodie. Inevitably also, the sharp curve reversals would cause lateral instability, and lead to a local buckling-type dislocation of microfibrils.

Initially such dislocations would appear only in isolated positions, within the S_2 layer. Then increasing overall deformation (creep) would cause stress concentrations in the region of the first of the dislocations. That would lead to transverse propagation of the effect into other more-resistant groups of microfibrils. Thus slip planes would propagate across the fibre wall, and later develop increased dislocation until they become more severe wall discontinuities. Such planes of microfibril folding (crinkling) also multiply along the length of the fibres, as illustrated by Dinwoodie (1968).

Comparable curvature reversals exist naturally in complete fibres and particularly at diversions of the fibres around rays; these may become the site of massive distortions which develop first across individual fibres, and then propagate across the walls of large groups of adjacent fibres. Finally that position of very severe distortions would be extended by stress concentration effects, to form the typical gross macroscopic failure zone, such as illustrated by Grossman and Wold (1971).

Overall, it appears that the progressive development of changes in microfibril outlines, and of slip planes and compression crinkles through the cell wall, may lead ultimately to formation of macroscopic failure zones. Such a sequence is the logical outcome of the simple interaction of microfibrils and matrix materials, when they are subjected to substantial imposed forces and to moisture changes, and possibly also temperature changes. At the same time, the series of such crinkles within the walls of the individual fibres must lead to a quite significant proportional shortening of the fibres, and of the wood tissue specimens.

Dinwoodie demonstrated that microscopic distortions increased with stress intensity during tests with short-duration loading. Similarly, the number and severity of the severe microfibril crinkles (referred to also as microscopic compression failures) were shown to increase with intensity of stress in beams drying under load over medium to long periods (Kingston and Armstrong, 1951; Armstrong and Kingston, 1962).

Referring to these effects in both compression specimens and beams, Armstrong and Kingston (1962) noted that in tests varying from 2 to 14 weeks, none of the very severe microscopic crinkles in the fibre walls 'were evident in beams maintained at constant moisture content under load'. However, they did not investigate the presence of slip planes. These account for only small contractions in the axial length of the microfibrils and the cell wall, in comparison to the contractions associated with 'microscopic compression failures'. Dinwoodie's study showed that the slip planes effectively are precursors of the more severe fibre-wall dislocations, and they are developed independently of involvement of moisture cycling under stress. Thus it seems reasonable to conclude that the relatively large magnitude of mechano-sorptive creep, in comparison to that of visco-elastic creep at similar imposed stresses, can be attributed to the greater numbers and severity of cell wall crinkles that are developed with mechano-sorptive creep.

Despite the very large deformation developed in mechano-sorptive creep, a big proportion of it is recoverable with moisture cycling after the loading force is removed (Armstrong and Kingston, 1962; Christensen, 1962). This is to be expected, partly because the creep-inducing compression of the matrix under load reduces the space for water, but with removal of the applied force, full swelling pressure can be generated in the matrix. Furthermore, the bonded positions are randomly distributed along microfibril lengths (Boyd and Foster, 1975). Consequently many microfibril lengths, which could be deflected substantially, but crinkled little by the loading, would bridge lengths of adjacent ones which were severely crinkled. That situation would facilitate substantial recovery, especially with moisture cycling in the absence of loading. Thus the anatomical model is compatible with the severe deflections in compression and bending, which can be developed by mechano-sorptive creep, and also with the recovery of most of it.

Development of failure in tension

In contrast to the substantial data available on the character of cell wall distortions which lead to failure in compression and bending, data relating to tension are rare. It was pointed out by Kollmann and Côté (1968), that for several reasons the potential tensile strength of microfibrils, fibres and specimens of wood generally could not be fully developed. Additionally, they noted that it is not practicable to make tensile tests of wood that provide precisely reliable strength data. On the other hand, it has been demonstrated that the strength of wood in tension could be up to three times that in compression.

To an increasing extent in recent years, it has been recognised that the structural performance of wood in tension is determined not by its apparent net strength (exclusive of macroscopically obvious weak zones), but by the strain development which causes a small defect to initiate a crack, and by the speed with which that crack propagates across the entire specimen. Furukawa (1978) investigated some aspects of crack development with notched microtome sections of Sugi wood. He noted that 'cracks always started at the tip of notches', then propagated at first relatively slowly, but later rapidly from there. The initial development was 'the incipient phase of failure'. He reported also that the characteristics of the failure zone in notched specimens were similar to those of specimens without notches.

Dinwoodie (1978) noted that microscopic slip planes and compression folds in wood fibres may develop in the standing tree, or during felling; in some cases even macroscopic failure zones develop at such times. From his observations of failure in tension, he stated that slip planes constitute weak links, and 'the line of crack propagation' was 'strongly influenced by the presence of slip planes and creases'. Because of the nature of the interactions between microfibrils and matrix in specimens subjected to pure tension (as previously discussed), it may appear that the extent of creep development (strain) would not have a significant effect on time to failure. On the other hand, as noted by Kollmann and Côté, pure tension is rare, as other modes of stress development tend to occur simultaneously and affect failure.

Both Furukawa and Dinwoodie observed shear failure lines between fibres, and between tension breaks across particular layers of the cell wall. Due to such weakness zones, tensile failure in long-duration loading, and especially in association with moisture cycling, could occur considerably sooner than might be anticipated from the stress intensity. Because of the high intensity of stress on the convex face, that effect would be more certain and more marked in respect of the tension side of bending specimens. Overall in respect of tensile loading, it is not the extent of strain which is the determinant of failure; it is the association of other stresses and positions of points and planes of weakness, which determine the initiation of a crack and its rate of propagation.

Similarly in respect of compressive forces, it is not the total strain associated with creep that determines the onset of failure. It is the development of distortion zones in microfibrils, as they are forced into new configurations, and also the presence of weaknesses such as slip planes and crinkles, which lead to elastic instability in the microfibrils and fibres, and to ultimate failure of the wood tissue. Because of the wide variability in the cell wall thickness and strength of adjacent fibres, even in a single tissue, and the additional wide variability of fibres within and between test specimens (Boyd, 1977a, 1980), the total strain at which critical instability (failure) will occur cannot be predicted. Effects of variations in environmental conditions, that are associated with loading the specimens, leads to failure having no relationship to any one suggested limit of sustainable strain. Undoubtedly however, the large deformations which develop in mechano-sorptive creep are likely to cause failures at much lower stresses than is the usual case with the relatively smaller deformations due to visco-elastic creep.

Creep in wood-fibre structural products

Both visco-elastic and mechano-sorptive creep have given rise to considerable research on reactions of hardboard and other wood-fibres boards, as noted by Grossman (1976). The major difference between those products and wood, is that processes in their manufacture lead to the fibres being incorporated in random orientations, instead of the parallel arrangement of fibres in wood. Hence, although the individual fibres, and their constituent microfibril and matrix constituents would react to stress (and also to moisture cycling and temperature changes) as already described, those reactions would be in accordance with the direction of that stress relative to each particular fibre.

The differing orientations of fibres would lead to corresponding differences in reaction to stress imposed in any one direction in the composite product. Accordingly, while the type of response to loading and moisture changes can be anticipated, predictions would be much more difficult on a quantitative basis. Additionally for such products, their physical responses may be influenced substantially by the nature of the bonding medium (glue, *etc.*) which envelopes the wood fibres, fibre bundles or flakes.

References

Alexandrov, W.G. & L.I. Djaparidze. 1927. Über das Entholzen und Verholzen der Zellhaut. Planta 4: 467-475.

Arima, T. 1972. Creep in process of temperature changes. 1. Creep in process of constant, elevated and decreased temperature. J. Japan Wood Res. Soc. 18: 349-353.

Armstrong, L.D. 1972. Deformation of wood in compression during moisture movement. Wood Sci. 5: 81-86.

Armstrong, L.D. & G.N. Christensen. 1961. Influence of moisture changes on deformation of wood under stress. Nature 191: 869-870.

Armstrong, L.D. & R.S.T. Kingston. 1960. Effect on moisture changes on creep in wood. Nature 185: 862-863.

Armstrong, L.D. & R.S.T. Kingston. 1962. The effect of moisture content changes on the deformation of wood under stress. Austr. J. Appl. Sci. 13: 257-276.

Bach, L. 1974. Rheological properties of beechwood in the ammonia-plasticized state. Mater. Sci. & Engng. 15: 211-220.

Balashov, V., R.D. Preston, G.W. Ripley & L.C. Spark. 1957. Structure and mechanical properties of vegetable fibres. 1. The influence of strain on the orientation of cellulose microfibrils in sisal leaf fibre. R. Soc. (Lond.) Proc. ser. B, 156: 460-468.

Bentum, A.L.K., W.A. Côté, A.C. Day & T.E. Timell. 1969. Distribution of lignin in normal and tension wood. Wood Sci. Technol. 3: 218-231.

Bethe, E. 1969. Festigkeitseigenschaften von Bauholz bei Lagerung im Wechselklima unter gleich-zeitiger mechanischer Belastung. Holz Roh u. Werkstoff 27: 291-303.

Bolton, A.J., P. Jardine, M.H. Vine & J.C.F. Walker. 1974. The swelling of wood under mechanical constraint. Holzforschung 28: 139-145.

Boyd, J.D. 1950. Three growth stresses. I. Growth stress evaluation. Austr. J. Sci. Res. 3: 270-293.

Boyd, J.D. 1972. Tree growth stresses. V. Evidence of an origin in differentiation and lignification. Wood Sci. Technol. 6: 251-262.

Boyd, J.D. 1977a. Relationship between fibre morphology and shrinkage of wood. Wood Sci. Technol. 11: 3-22.

Boyd, J.D. 1977b. Interpretation of X-ray diffractograms of wood for assessments of microfibril angle in fibre cell walls. Wood Sci. Technol. 11: 93-114.

Boyd, J.D. 1980a. Relationship between fibre morphology, growth strains and physical properties of wood. Austr. For. Res. 10: 337-360.

Boyd, J.D. 1980b. Biophysical controls of cellulose formation. In: C.H.A. Little (ed.), Control of Shoot Growth in Trees: 184-236. Proc. IUFRO Workshop on Xylem and Shoot Physiology, Fredericton, New Brunswick, Canada.

Boyd, J.D. 1982. Biophysics of microfibril orientation in plant cell walls. In preparation.

Boyd, J.D. & R.C. Foster. 1974. Tracheid anatomy changes as responses to changing structural requirements of the tree. Wood Sci. Technol. 8: 91-105.

Boyd, J.D. & R.C. Foster. 1975. Microfibrils in primary and secondary wall growth develop trellis configurations. Canad. J. Bot. 53: 2687-2701.

Chen, M.M. 1974. A proposed explanation for the phenomenological rheology of prefrozen redwood. Wood Sci. 7: 34-42.

Choong, E.T. 1969. Effect of extractives on shrinkage and other hygroscopic properties of ten southern pine woods. Wood and Fiber 1: 124-133.

Christensen, G.N. 1962. The use of small specimens for studying the effect of moisture content changes on the deformation of wood under load. Austr. J. Appl. Sci. 13: 242-256.

Cleland, R. 1971. The mechanical behaviour of isolated Avena coleoptile walls subjected to constant stress. Properties and relation to cell elongation. Plant Physiol. 47: 805-811.

Davidson, R.W. 1962. The influence of temperature on creep in wood. For. Prod. J. 12: 377-381.

Dinwoodie, J.M. 1968. Failure in timber. 1. Microscopic changes in cell-wall structure associated with compression failure. J. Inst. Wood Sci. 21: 37-53.

Dinwoodie, J.M. 1978. Failure in timber. 3. The effect of longitudinal compression on some mechanical properties. Wood Sci. Technol. 12: 271-285.

Erickson, R.W. 1968. Drying of prefrozen redwood – fundamental and applied considerations. For. Prod. J. 18 (6): 49-56.

Erickson, R.W., M.M. Chen & T. Lehtinen. 1972. The effect of unidirectional diffusion and pre-freezing upon flexural creep in redwood. For. Prod. J. 22: 56-60.

Erickson, R.W. & D.J. Sauer. 1969. Flexural creep behaviour of redwood heartwood drying from the green state. For. Prod. J. 19 (12): 45-51.

Eshbach, O.W. 1952. Handbook of engineering fundamentals. 2nd Ed. John Wiley & Sons, New York; Chapman & Hall, London.

Frei, E. & R.D. Preston. 1961. Cell wall organization and wall growth in the filamentous green algae Cladophora and Chaetomorpha. I. The basic structure and its formation. R. Soc. (Lond.) Proc. ser. B, 154: 70-94.

Frey, A. 1926. Submikroskopische Struktur der Zellmembranen. Eine polarisationsoptische Studie zum Nachweis der Richtigkeit der Micellartheorie. Jahrb. Wiss. Bot. 65: 195-223.

Frey-Wyssling, A. 1976. The plant cell wall. Gebr. Borntraeger, Berlin/Stuttgart.

Fujita, S. & A. Takahashi. 1969. Rheological properties of tropical wood. II. On the histological effects in mechanical properties of tropical wood applied to stress and temperature during drying. J. Japan Wood Res. Soc. 15: 271-277.

Furukawa, I. 1978. Optical microscopic studies on the longitudinal tensile failure of notched microtome sections. J. Japan Wood Res. Soc. 24: 598-604.

Gardner, R., E.J. Gibson & R.A. Laidlaw. 1969. Effects of organic vapours on the swelling of wood and on its deformation under load. For. Prod. J. 17: 50-51.

Grossman, P.U.A. 1953. The recovery of plywood after compression at elevated temperatures. Austr. J. Appl. Sci. 4: 98-106.

Grossman, P.U.A. 1976. Requirements for a model that exhibits mechano-sorptive behaviour. Wood Sci. Technol. 10: 163-168.

Grossman, P.U.A. 1978. Mechano-sorptive behaviour. In: General constitutive relations for wood and wood-based materials. Report of Workshop sponsored by The Nat. Sci. Found.; Eng. Mechanics Sec.; Solid Mechanics Program: 313-322.

Grossman, P.U.A. & M.B. Wold. 1971. Compression failure of wood across the grain. Wood Sci. Technol. 5: 147-156.

Grozdits, G.A. & G. Ifju. 1969. Development of tensile strength and related properties in differentiating coniferous xylem. Wood Sci. 1: 137-147.

Harada, H. 1965. Ultrastructure and organization of gymnosperm cell walls. In: W.A. Côté (ed.), Cellular Ultrastructure of Woody Plants: 215-233. Syracuse Univ. Press, Syracuse, N.Y.

Hearmon, R.F.S. & J.M. Paton. 1964. Moisture content changes and creep in wood. For. Prod. J. 14: 357-359.

Hill, R.L. 1967. The creep behaviour of individual pulp fibres under tensile stress. TAPPI 50: 357-379.

Jacobs, M.R. 1938. The fibre tension of woody stems, with special reference to the genus Eucalyptus. Commonw. For. Bur. Austr. Bull. No. 22.

Kelsey, K.E. 1963. A critical review of the relationship between the shrinkage and structure of wood. CSIRO, Austr. Div. For. Prod. Technol. Paper No. 28.

Kerr, A.J. & D.A.I. Goring. 1975. The ultrastructural arrangement of the wood cell wall. Cellulose Chem. Technol. 9: 563-573.

Kingston, R.S.T. 1962. Creep relaxation and failure in wood. Research 15: 164-170.

Kingston, R.S.T. & L.D. Armstrong. 1951. Creep in initially green wooden beams. Austr. J. Appl. Sci. 2: 306-325.

Kingston, R.S.T. & B. Budgen. 1972. Some aspects of the rheological behaviour of wood. IV. Nonlinear behaviour at high stresses in bending and compression. Wood Sci. Technol. 6: 230-238.

Kisser, J. & H. Frenzel. 1950. Mikroskopische Veränderungen der Holzstruktur bei mechanischer Überbeanspruchung von Holz in der Faserrichtung. Schr. Reihe öst. Ges. Holzforsch. 2: 7-31.

Kitahara, K. & K. Yukawa. 1964. The influence of change in temperature on creep in bending. J. Japan Wood Res. Soc. 10: 169-175.

Kollmann, F.F.P. & W.A. Côté. 1968. Principles of wood science and technology. 1. Solid wood. Springer, Berlin/Heidelberg/New York.

Kubler, H. 1973. Hygrothermal recovery under stress and release of inelastic strain. Wood Sci. 6: 78-86.

Leaderman, H. 1944. Elastic and creep properties of filamentous materials and other high polymers. Textile Foundation, Washington.

Mark, H. 1952. Cellulose: Physical evidence regarding its composition. In: L.E. Wise & E.C. Jahn (eds.), Wood Chemistry. Vol. 1: 132-seq. Reinhold, New York.

222

Mark, R.E. & P.P. Gillis. 1973. The relationship between fiber modulus and the S_2 angle. TAPPI 56: 164-167.

Murphey, W.K. 1963. Cell wall crystallinity as a function of tensile strain. For. Prod. J. 13: 151-155.

Nearn, W.T. 1955. Effect of water soluble extractives on the volumetric shrinkage and equilibrium moisture content of eleven tropical and domestic woods. Penn. State Univ., Agr. Exp. Sta. Bull. No. 598.

Nicholson, J.E. 1973. Growth stress differences in eucalypts. For. Sci. 19: 169-174.

Onaka, F. 1949. Studies on compression and tension wood. Wood Res. Inst., Wood Res., Kyoto, Bull. No. 1.

Pearson, R.G., N.H. Kloot & J.D. Boyd. 1958. Timber engineering design handbook. CSIRO & Melbourne Univ. Press, Victoria, Australia.

Preston, R.D. 1941. The fine structure of phloem fibres. II. Untreated and swollen jute. R. Soc. (Lond.) Proc. ser. B, 130: 103-112.

Preston, R.D. 1974. The physical biology of plant cell walls. Chapman & Hall, London.

Preston, R.D. & M. Middlebrook. 1949. The fine structure of sisal fibres. J. Text. Inst. 40: T715-T722.

Resch, H. & B.A. Ecklund. 1964. Permeability of wood − exemplified by measurements on redwood. For. Prod. J. 14: 199-206.

Roelofsen, P.A. 1959. The plant cell wall. Gebr. Borntraeger, Berlin-Nikolassee.

Scallan, A.M. 1974. The structure of the cell wall of wood − a consequence of anisotropic inter-microfibrillar bonding? Wood Sci. 6: 266-271.

Schniewind, A.P. 1966. On the influence of moisture content changes on the creep of beech wood perpendicular to the grain including effects of temperature and temperature change. Holz Roh u. Werkstoff 24: 87-98.

Schniewind, A.P. 1967. Creep rupture life of Douglas fir under cyclic environmental conditions. Wood Sci. Technol. 1: 278-288.

Schniewind, A.P. 1968. Recent progress in the study of rheology of wood. Wood Sci. Technol. 2: 188-206.

Simpson, W.T. 1971. Moisture changes induced in red oak by transverse stress. Wood and Fiber 5: 13-21.

Stamm, A.J. 1964. Wood and cellulose science. Ronald Press, New York.

Tieman, H.D. 1944. Wood technology. 2nd Ed. Pitman Publ., New York.

Timoshenko, S. 1940. Strength of materials. 1. Elementary theory and problems. 2nd. Ed. Van Nostrand, New York.

Tokumoto, M. 1973. Moisture recovery of drying set. II. Effect of quantity of adsorbed moisture and dry-wet cyclings on set recovery in wood. J. Japan Wood Res. Soc. 19: 585-591.

Urakami, H. & K. Nakato. 1966. The effect of temperature on torsional stress relaxation of wet Hinoki wood. J. Japan Wood Res. Soc. 12: 118-123.

Wardrop, A.B. 1971. Lignin occurrence and formation in plants. In: K.V. Sarkanen & C.H. Ludwig (eds.), Lignins: 19-41. John Wiley & Sons, New York.

Wood, L.W. 1951. Relation of strength of wood to duration of load. U.S. For. Prod. Lab. Report R1916.

Yamada, T., K. Sumiya & N. Kanaya. 1966. Rheo-optics of wood. 1. Variations of infra-red spectra with creep in Hinoki (Chamaecyparis obtusa Sieb. et Zucc.). Wood Res. Bull., Wood Res. Inst., Kyoto Univ. No. 38: 21-31.

The application of statistics and computing in wood anatomy*

J. BURLEY and R.B. MILLER

Department of Agricultural and Forest Sciences, Oxford University, Commonwealth Forestry Institute, South Parks Road, Oxford, England, and Center for Wood Anatomy Research, Forest Products Laboratory, U.S.D.A. Forest Service, Madison, Wisconsin 53705, U.S.A.

Summary: In published anatomical research up to the 1960 decade there has been little application of statistical method other than presentation of mean values. With the recent expansion of statistical teaching and the wider availability of computers, statistical treatment of anatomical data should increase. This paper presents a checklist of univariate and multivariate statistical techniques relevant to identifying and characterising sources and patterns of variation of anatomical features within and between trees and geographic locations. Non-statistical applications of computers are described including the curatorial management of wood collections (xylaria) and the identification of wood samples. Attention is drawn to the IAWA 'Standard list of characters suitable for computerized hardwood identification'.

Introduction

In the thirty years that have passed since the first edition of 'Anatomy of the Dicotyledons' (Metcalfe and Chalk, 1950) was published, two scientific developments stand out as having major impacts on anatomical research. The first of these is the electron microscope, particularly the scanning electron microscope of which the applications (in conjunction with hand lens and optical microscope) were described by Cutler (1979) in the first volume of the revised edition (Metcalfe and Chalk, 1979).

The second is the computer which, at first sight, may appear to have little to do with wood anatomy but which, with its increasing availability, sophistication and acceptability, offers support for a wide range of activities related to studies of wood anatomy. Among the many benefits of the widespread availability of pocket, desk-top and main-frame computers is the wider application by biologists of statistical methods in designing and evaluating their research,

---

* This paper is a much enlarged version of a chapter in the second edition of 'Anatomy of the Dicotyledons' (Metcalfe and Chalk, 1982).

although it must be admitted that the use of packaged programs that are commonly provided with the machines does not ensure the correct use or understanding of statistical method. Further, packaged programs often do not make the fullest use of data nor do they always provide results in a form easily intelligible to or desired by the researcher; user-oriented programs are preferable and it is encouraging to note the expansion of teaching of computer programming to scientists. In this chapter we do not discuss the many modern applications of computers in interfacing between laboratory equipment and automatic systems of recording and manipulating data.

For many of the anatomical characteristics studied by wood anatomists and technologists considerable variation exists between plant taxa, populations and individuals; these may be ascribed to genetic and environmental effects. Further there is commonly variation within individuals attributable to ontogenetic and environmental effects. Environment here includes not only the natural features of biotic, climatic and edaphic factors but also the artificial features of managed plantations.

If variation exists, more than one sample is required to estimate the mean for a tree, taxon, site or management treatment; statistical method is required to give known confidence limits in:

1) the description of variation of one anatomical characteristic,
2) the correlation between pairs or groups of anatomical characteristics,
3) the correlation between anatomical characteristics and environmental factors,
4) the identification or classification of wood samples.

In much of the published anatomical work up to the early 1960s there was little use of statistics other than presentations of mean values. This reticence on the part of botanists, and indeed of many other biologists, was due to lack of statistical training, to lack of appreciation of the need for rigorous treatment of data and to lack of contact with professional statisticians or biometricians. Now, with increasing university teaching of statistics to biologists and with growing numbers of readable and relevant text-books, statistical analyses and stars of significance abound, often merely to satisfy the desires of journal editors. Clearly a balanced view and use of statistics is necessary and the advice of the professional should always be sought.

This chapter is not intended as a text-book of statistics but rather as a checklist of methods to be considered in analysing data. Similarly it is not a computing manual but a summary of applications of computers to anatomical problems and activities.

Statistical methods

Collection and checking of data

Depending on the objectives of research and on the equipment used, data may be recorded on hand-written paper sheets or cards, hand-punched cards or paper tapes, mark-sensing cards, floppy disks or magnetic tapes; some of these may be automatic recordings of data or voice. Whichever method is used all data should be checked as soon as possible for obvious recording errors; they should be checked again after transcription into the permanent form of record (office file or computer store). If possible, initial, independent checking should be done while the wood samples or slides are still nearby and the staff concerned are still familiar with them so that samples yielding anomalous data can be re-examined.

Preliminary sampling

Excluding cases of very small populations (*e.g.* the assessment of a single characteristic, such as tree height, in a fixed number of trees, as in a small arboretum plot), it is usually neither practicable nor necessary to measure all the individual samples possible (*e.g.* all fibres in all trees of a given species); thus sampling is required and, ideally, a preliminary sample is taken to determine the number and distribution of samples necessary in the field and in the laboratory to achieve acceptable precision. Various techniques for determining sample size are available (see *e.g.* Freese, 1967; Snedecor and Cochran, 1967) but the best method of allocating future sampling resources to various hierarchical levels of samples having random effects involves the use of variance components (see *e.g.* Bliss, 1967; Burley *et al.*, 1970). A further complication arises when systematic patterns of variation exist as, for example, in the radial or axial variation within trees of fibre and vessel characters; systematic sampling may then be necessary with a different allocation of random sub-samples within each systematic position.

Initial statistical examination of data

Biological data are often analysed (*e.g.* by analysis of variance) without preliminary tests that should, stricktly, be applied to all data before analysis but particularly when new types of data become available. These tests include examination of normality of distribution, skewness and kurtosis, and tests of homogeneity of variance. The effects of non-normality or heterogeneity vary with the type of analysis undertaken subsequently but they are often overlooked. Where such effects are detected the data should be transformed if possible by an appropriate transformation (*e.g.* angular, logarithmic, square root, probit or reciprocal – see Bliss, 1967; Jeffers, 1959). Alternatively, non-parametric (distribution-free) methods of analysis may be used.

Before proceeding to other numerical analyses it is usually worthwhile to examine the distribution of the data graphically, particularly where the object of the research is to determine patterns of anatomical variation within individual trees or between geographic locations. For the simple categorisation of a single taxon, however, a simple statement of mean and standard error may be sufficient.

Characterisation of patterns of univariate variation

Within trees. – Since the proposition of Sanio's laws for variation of tracheid size in conifers (Sanio, 1872; Bailey and Shephard, 1915), there have been many reports of systematic patterns of variation within individual trees of both conifers and dicotyledons. Not until about 1970, however, were attempts made to describe such patterns mathematically (Burley, 1969, 1970; Burley and Andrew, 1970; Andrew and Burley, 1972).

This characterisation of variation within a single tree requires linear and non-linear regression of each character in relation to single or multiple independent positional variables such as the height or radial location of each sample. The mathematics of regression analysis is also used to examine the correlation between pairs of traits.

For each individual trait there is also the problem of estimating a meaningful average value for each tree in order to compare or combine data from several trees. Such tree means can be estimated by integration of polynomial curves or by a weighting process in which the data for a particular location in the tree are weighted by the proportional cross-sectional area or whole tree volume that they represent. Where tree growth is in the form of approximately concentric but unequal annuli or sheaths, simple arithmetic averaging is not appropriate.

Between trees. – To the delight of tree breeders but to the despair of wood anatomists, technologists and users, trees vary in their mean wood characters. Such variation is detected by the analysis of variance (see *e.g.* Snedecor and Cochran, 1967; Scheffé, 1959) which indicates the ratios between the variances of means and an appropriate residual variance (usually mis-named the 'error'). These ratios are used to determine the statistical significance of differences between trees, *i.e.*, they allow the researcher to reject the null hypothesis at a given level of probability. Many of the published papers and doctoral theses of Zobel, co-workers and students demonstrate such tree-to-tree variation (see *e.g.* Zobel *et al.*, 1972).

Between environments. − The analysis of variance may be extended, using an appropriate fixed, random or mixed mathematical model, to examine differences between environments (*e.g.* between natural populations of a geographically widespread species or between sites or managerial treatments for a planted species, and of course between different taxa − see Hans and Burley, 1972; Hans *et al.*, 1972a, b). When significant differences between means are detected, correlation and regression analyses are again used to examine the relationship of the measured variable to some independent variable (such as site altitude or latitude − see Van der Graaff and Baas, 1974; Van den Oever *et al.*, 1981) for description or prediction.

Standard errors and coefficients of variation. − When the variance ratios indicate that a statistically significant effect is attributable to environments, trees, positions or some other factor of interest, the standard errors of means or differences should be examined to see if the differences are statistically, biologically or economically important. Coefficients of variatation indicate the extent of variation in relation to the overall mean; they may be calculated in different ways as the ratio of the standard error to the mean or as the ratio of a within-sample standard deviation to the mean. The latter is the traditional definition but in either case the coefficient is sensitive to errors in the estimate of the mean and has rather limited use, particularly in the analysis of attributes or of non-linearly transformed data.

Range tests. − If important differences are found among a group of treatments, multiple range tests (*e.g.* Studentized, Duncan, Scheffé) can be used to test specific hypotheses or to group the means into sets within each of which there is no statistically significant variation. However, the unwary user must be warned of the intrinsic unsoundness of such tests. (See *e.g.* Snedecor and Cochran, 1967, p. 274.)

Variance components. − The analysis of variance of data from such studies often stops at the calculation of means, variance ratios and their associated significance levels. However, the expectation mean squares from the appropriate linear model allow the calculation of variance components. These, particularly if expressed as a percentage of total variation, indicate the distribution of variation among all sources and hence their relative importance. Additionally variance components are used to estimate population genetic parameters and heritabilities.

Correlation between traits. – In addition to determining the sources of variation in individual characteristics it is valuable to know the correlations between pairs of traits. When several traits are measured in a given experiment the matrix of correlation coefficients for all possible pairs quickly indicates those that are significantly correlated. Analogous with the analysis of variance, an analysis of covariance allows estimation of phenotypic, genetic and environmental correlations. In many studies it is also necessary to know the correlations between anatomical characters and end-use properties of wood.

Multivariate analysis. – Biologists are now becoming familiar with the concepts and methods of univariate analysis of variance and with techniques such as correlation analysis (between two measured variables) and multiple regression analysis (in which variation in one measured, dependent variable is explained by joint variation in several, error-free, independent variables). However, there is some reluctance to use the more complex multivariate methods that are appropriate to obtain more information when several measurements are made on one sample (although it should be noted that much of the statistical theory and practice was developed in response to genetic and taxonomic problems).

To some extent this can be attributed to lack of training, to mathematical complexity and to lack of computing facility, but it is also due in part to confusion in the literature about the names and functions of the different multivariate techniques. The object of this chapter is not to reproduce the statistical theory and algebra (which are described in many textbooks including Anderson, 1958; Seal, 1964; and Kendall, 1975) but to indicate, with examples and sources where appropriate, the types of analysis that may be applicable to studies of wood anatomy. It must be recognised that some types of analysis are only generalisations of others; *e.g.* canonical correlation analysis is a generalisation of multiple regression analysis and a two-group discriminant analysis is a special case of canonical analysis of discriminance.

The objectives of multivariate methods include (1) reduction of the dimensionality of the original data and assistance with biological interpretation (all methods), (2) the generation of hypotheses (principal components analysis), (3) correlations between two sets of multiple variables (canonical analysis), (4) classification *sensu lato* or identification, *i.e.* allocation of a sample to one of a previously defined set of groups (discriminant analysis), and (5) classification *sensu stricto, i.e.* generation of a classification system (cluster analysis). For most practical purposes multivariate normal distributions of data are assumed.

Principal components analysis. — This type of analysis (PCA) is a technique for reducing the dimensionality of a multivariate set of data; technically it is the calculation of the principal axis transformations of quadratic surfaces in a multi-dimensional space. The components are linear combinations of the raw variables found as eigenvectors of the covariance or correlation matrix; they are orthogonal (*i.e.* axes at right angles) and each accounts in turn for the greatest amount of residual variation. No assumptions about the distributions are required but scale effects can be important although they can be eliminated easily. No confidence limits can be calculated.

The practical result is to reduce the many features that anatomists commonly assess to a few new variables that are mathematical artefacts yet that can have biological interpretation and may assist in formulating biological or technological hypotheses. For each original individual sample, scores based on these components can be calculated, and these can be examined with regard to grouping of samples in relation to origin, treatment, species, *etc.*

Examples of immediate interest to wood anatomists are the studies by Davidson (1972), in which four principal components explained 93 percent of the total variation in 15 properties of trees of *Eucalyptus deglupta* Blume, and by Cailliez and Gueneau (1972), in which two principal components explained 84 percent of the variation among nine technological properties in 414 Malgasian trees from 13 genera, 125 species and 22 ecological areas. Other relevant studies are those of Jeffers (1962, 1967a), Van den Oever *et al.* (1981), and Ross and Morris (1971).

Factor analysis. — This term is commonly and loosely applied as a collective name for multivariate techniques but strictly it is a definite form. (Seal, 1964, suggested the use of capital initial letters FA to identify this specific type of Factor Analysis.) Like PCA, FA is a statistical technique for reducing a large number of observed, correlated variables to a small number of abstract uncorrelated variables; however, FA is distinguished from PCA by at least one feature.

Each of the original variables is supposed to be analysed into a small number of mutually uncorrelated common factors with a unique uncorrelated residual that is also not correlated with any of the remaining variates; this assumption is unlikely to be met strictly in anatomical studies where common genetic history, cell ontogeny from common ancestral cells, systematic ageing and positional effects, and common environment are likely to cause considerable correlation between features.

For both FA and PCA, the orthogonal axes of common factors may be rotated to new orthogonal or oblique axes to agree with some preconceived model hypothesis. Thus while PCA can be hypothesis-building, FA is used to clarify the biological significance of results. A large number of published biological examples of FA were discussed by Seal (1964).

Canonical analysis. – The two preceding analyses (PCA and FA) were concerned with summarising information derived from many samples, each assessed for many features. The following sections are concerned with comparing or arranging the samples themselves in a biologically meaningful manner, taking into account the variability within and between samples. Unlike PCA and FA, scale effects are unimportant.

The objective of canonical analysis (CA) is to identify linear combinations of two sets of variables, maximising their correlation. The two sets must be differentiated *a priori* and not *a posteriori* after examining the correlation matrix. Canonical analysis includes both canonical correlation analysis (CCA), in which the sample is unstructured within both sets of variables, and canonical analysis of discriminance (CDA), in which one set (only) identifies groups in the sample by using constant values (such as group means) within a group. A typical example would be the data from 20 anatomical features and six climatic or edaphic features in a study of geographic variation in wood anatomy. Scores for each original sample can be plotted for each pair of canonical variates to examine geographic or taxonomic relationships. See, for example, the relationship of *Pinus kesiya* needle anatomy to geographic seed source in Burley and Burrows (1972) and the relation of *Betula lutea* Michx. leaf variation to source environment in Dancik and Barnes (1975). Canonical analysis has been extended to deal with more than two sets of variables at a time (Kettenring, 1971) and a special form of canonical correlation analysis is multiple discriminant analysis.

Where the group-identifying variables are mutually orthogonal dummy variables, the results of CDA are effectively identical to those of multiple discriminant analysis (MDA). It is noteworthy that when both sets of variables consist of group means the analysis reverts to CCA. Falkenhagen and Nash (1978), in classifying provenances of *Picea sitchensis* (Bong.) Carr., compared CCA with MDA (which they referred to as canonical variate analysis or discriminant analysis).

Multiple discriminant analysis. – This type of analysis (MDA) is concerned with samples of which the members belong to one of two or more *a priori* groups or classes and it determines linear combinations of variables that provide discriminant functions and distances among the groups, maximising the differences between groups. The discriminant functions are mutually orthogonal and account for as much as possible of the remaining variation between classes and are canonical variables in the broad sense. The objective is to provide a function that will enable the placing of a new, unknown sample into the appropriate class based on its own discriminant function score. Data must be normally distributed but a generalised distance function is independent of the scale chosen and is thus suitable when the variables are principal components.

The first use of discriminant analysis with woody plants was by Hopp (1941) in describing the various forms of shipmast locust (*Robinia pseudoacacia* L.) and Mergen *et al.* (1966) used the technique to examine hybrids between *Eucalyptus cinerea* F.v.M. and *E. maculosa* R.T. Baker. The paper by Rouvier (1966), although concerned with animals rather than plants, gave a useful comparison of PCA with MDA; it showed that canonical variables can be expressed as functions of principal components. Mariaux and Vitalis-Brun (1979) used principal components analysis and factorial discriminant analysis of anatomical features in identification of 20 closely related genera of Sapotaceae.

Cluster analysis. − When well defined classes exist MDA permits the allocation of an unknown individual to one of the classes; this is essentially identification. When the observations on the individual samples are themselves to be used to suggest a classification scheme, this is termed classification or cluster analysis. Many techniques exist for the detection and portrayal of pattern among biological entities, commonly represented in a hierarchical dendrogram.

For statistical clustering, variables must be normally distributed and the analysis is not independent of scale, except in the case where the correlation coefficient is used to define the distance between points in multidimensional space. Non-statistical, geometric clustering based on similarity matrices of categorical information is commonly used in numerical taxonomy (see Sokal and Sneath, 1963) with single linkage, complete linkage, average linkage, centroid clustering and other methods. Hogeweg and Koek-Noorman (1975), in their study of wood anatomy in some groups of the Rubiaceae, contested the claim of Sokal and Sneath (1963) that such methods would eliminate the need for character selection and character weighting in taxonomy. (Cluster analysis was used also by Koek-Noorman and Hogeweg (1974) in describing the wood anatomy of other tribes of the Rubiaceae.) More detailed comparisons and contrasts of identification and classification were given by Reyment (1969) and Gower (1975).

Concluding comments on statistics

The foregoing should be treated as a guide to the types of statistical treatment that can be considered in designing surveys or experiments and in analysing and interpreting the resulting data. Appropriate statistical rigour is always necessary and multivariate methods can simplify the interpretation and presentation of multiple observations; however, they can not replace, but must be used as an adjunct to, skilled observation and judgement by the professional anatomist and taxomist.

Non-statistical applications of computers to wood anatomy

The initial advantage of computers in wood anatomy clearly lies in their ability to perform tedious and complex statistical calculations and manipulations quickly and precisely. However, two other applications have potential interest for wood anatomists, particularly those concerned with the curation of wood collections and the identification of wood samples. We exclude here any detailed consideration of those uses that are now routine in administration such as the maintenance of address lists, printing of adhesive mailing labels, inventory and payroll accounting, *etc.,* which are relevant to all scientists and institutions.

We also avoid detailed discussion of the rapidly expanding computer systems of storing and retrieving bibliographic information in which published reports may be retrieved by any combination of authors, dates title, keywords and journals. Of such systems perhaps the Californian Lockheed DIALOG system has the broadest coverage and appeal to anatomists; it can determine numbers of references meeting the multiple selection criteria, list their titles and print their abstracts to a remote telephone terminal anywhere in the world.

Information retrieval and management of wood collections

All wood collections have some system whereby information and data regarding the wood specimens can be retrieved. Some systems give each wood specimen an accession number and the pertinent information. Scientific name, collector and number, origin and location of any voucher specimens are placed on cards or in a record book together with the accession number. Duplicate sets are often made and filed alphabetically by genus and/or by family. The wood samples may be filed by accession number or alphabetically by family. Other wood collections attach all the information to the wood specimen and then file the wood specimen alphabetically by family; generally, no accession numbers are assigned. There are definite variations from one wood collection to another but essentially the systems are similar and all the information known about the specimen is filed in one place. No computerised collection management or information retrieval programs are now used to help curators manage their collections.

The advantage of a computerised collection management system is that much more information can be retrieved. With a card system even as elaborate as the one with the Samuel James Record Memorial Wood Collection (SJRw) housed at the U.S. Forest Products Laboratory, Madison, Wisconsin, only a limited amount of data can be retrieved. This collection has one set of cards arranged numerically by accession number, one alphabetically by genus and then by the specific epithet, one alphabetically by family, genus, and specific epithet, and one alphabetically by common name. In addition there is a file containing the original lists of specimens from the collector or institution arranged alphabetically by country of origin. This system of cards enables the curator to retrieve much information including what specimens were collected in individual countries and who the collectors were. Information not easily retrieved includes lists of vouchered and type specimens, and locations of voucher specimens, comprehensive lists of collectors and/or specimens, specimens collected in a particular state or province, lists of specimens for which there are microscope slides, and lists of misidentified and unidentified specimens. When trying to update name changes from comprehensive flora studies, it is difficult and time consuming to make the various changes on four sets of cards, the original list, and sometimes the herbarium sheet. Generally, the change and source is noted on each card which must first be located in the file, changed, and then refiled. With a computer system there would be some timesaving but, more importantly, information that was not retrievable with cards would be retrievable with a computerised system.

Some type of computer is in practically every laboratory and institution, and most scientists have access to a large computer either directly or via the telephone and interactive terminal. With this increase in speed and memory storage scientists soon developed programs for information retrieval. At first these programs were simple and straightforward programs mainly for such things as bibliographies. Today, complex sophisticated information retrieval programs coupled with large computers with great quantities of memory are available. Museums, taxonomists, systematists and others with large collections of objects wrote, developed, and adapted information retrieval systems for their particular needs. SELGEM, GRIPHOS, MCN, MIRS, and TAXIR (EXIR) are the acronyms of just a few systems that are available and in use in some museums. There are books, pamphlets, catalogues, and articles on these systems, and in some cases workshops for their implementation and use. The ASC (Association of Systematics Collections) was formed in 1973 to 'foster care, management, preservation, and improvement of systematics collections.' This organisation has done much to promote collection management via electronic data processing through their newsletter, bulletins, and other publications. To date no wood collection has implemented electronic data processing for collection management but the information and assistance is available. Once one collection adopts a system, other institutions may find it feasible to use the same system and format.

Although there are many computerised systems available that can accomplish a wide variety of functions and are suitable for curatorial duties in a xylarium, *TAXIR* and *SELGEM* are two that are widely available and have been adopted by some institutions. *TAXIR (TAXonomic Information Retrieval*, now replaced by *EXIR, EXecutive Information Retrieval*) is designed to be used in taxonomic work of all types. It is a general purpose system useful for any type of data or description of either specimens or taxa. *TAXIR* can keep efficient records on loans and accessions but can also be used for the purpose of recording all types of classificatory information. *TAXIR* is written in Fortran IV and thus it is relatively easy to convert from one computer to another. Also, because of its modular construction not all the system needs to be converted. This modular concept also allows the user to adapt the system to a smaller computer. The user's manual for *TAXIR* is available from the University Computer Center, University of Michigan, Ann Arbor, Michigan 48109, U.S.A. Information and the user's manual for *EXIR* are available from the Information Sciences/Genetic Resources Program, University of Colorado, Boulder, Colorado, U.S.A.

SELGEM (SELf-GEnerating Master) is a package of generalised computer programs for general information processing, collection documentation, and for related research-oriented projects. *SELGEM* is written in COBOL and is presently operational on a number of different computers. There are user manuals, workshops, and a newsletter called MESH to assist institutes in establishing, maintaining, and modifying this system. Further information about *SELGEM* is available from the National Museum of Natural History, Washington, D.C. 20560, U.S.A.

TAXIR and *SELGEM* are capable of handling much more than just the management requirement of even the largest wood collection. Both offer the capabilities of adding and retrieving such information as physical and mechanical properties or macro- and microscopic anatomical characteristics, taxonomic history, synonymy for each specimen, species, or genus. The possibilities are almost limitless, but the cost may prove to be prohibitive. Any institution contemplating the adoption of a large sophisticated information retrieval program should consider the technological possibilities that may develop in the near future. Generally, wood collections contain fewer than 30,000 specimens and the possibility of the collection increasing by 5,000 specimens per year is remote. A minicomputer, coupled with the storage capacity available on floppy disks, might be the answer for a wood collection. This type of system is rather limited in memory capacity and the searches for various data may be lengthy but the overall cost is lower. Perhaps in many wood collections the non-computerised file card system is still the most efficient method for collection management.

Wood identification

Many anatomists and curators of wood collections are called on to identify wood samples. Experienced workers may be able to identify many samples immediately or with little effort; for less experienced researchers and for unfamiliar species a long process of describing many characters and comparing them with known samples is necessary.

In the mid 1940s two scientists independently recognised that the mechanical devices peripheral to computers could be an aid in wood identification. Dr. W.W. Varossieau at the Forest Products Research Institute T.N.O. in Delft, the Netherlands, and Dr. B.F. Kukachka at the U.S. Forest Products Laboratory in Madison, Wisconsin, U.S.A., independently developed systems of identification based on computer cards and a card sorting machine. Essentially, both methods adopted the principles of the multiple-entry, marginally perforated cards which were developed by Clarke (1938).

A number of different identification systems were proposed prior to 1938 (consult Pfeiffer and Varossieau, 1945; and Varossieau, 1948), but Clarke's multiple-entry card system was a major step forward in wood identification. Soon, marginally perforated cards were developed by such wood anatomists as Phillips (1941) in England, Beversluis (1943) in the Netherlands, Normand (1946) in France, Dadswell and Eckersley (1941) in Australia, Thunell and Perem (1947) in Sweden, Corothie (1948) in Venezuela, and others. Later, still others were developed and revisions of the established cards were proposed and adopted (Balan Menon, 1957; Gottwald, 1958; Kribs, 1959; Kukachka, 1960; Brazier and Franklin, 1961; Sudo, 1963; Normand, 1972). Each system developed its own set of characters, character states, and arrangement, but the principle was the same. Each card represented one taxon, generally a species or genus. Initially, the cards had holes evenly spaced along the margins and each hole represented a character or character state. If a hole were opened or slotted using a specially designed punch or clipper, that character or character state was positive. Negative character states were represented by an unclipped hole. To identify an unknown, the state of any character was determined and then a rod was pushed through all the cards for the hole representing that character state. With the rod in the hole, all cards were lifted. Those with an open slot (positive) fell and those with a hole (negative) remained on the rod. This process continued until only one taxon remained. The biggest advantage of this system is the multiple-entry capabilities, *i.e.* a polyclave. The operator selects any character whereas a dichotomous key dictates the order of character selection, *i.e.* a monoclave. The disadvantages of the marginally perforated card systems are the limited number of characters that can be placed around the margins of the cards, the deterioration of the cards, and the limited number of taxa (cards) that can be conveniently manipulated. The biggest complaint and frustration, however, is the failure of all the cards to fall out when they should. Varossieau and Kukachka solved this problem by substituting a card sorting machine for the rod.

Varossieau (1948) converted the Dutch decimal method earlier proposed by Pfeiffer and Varossieau (1945) to an 80-column card known as the Hollerith System. About the same time Kukachka developed a system using an 80-column IBM card. At the International Association of Wood Anatomists (I.A.W.A.) meetings in Stockholm in July, 1950, Varossieau and Kukachka presented a joint paper entitled 'Suggestions for a standardized method of identification with mechanically operated punched cards.' Eighteen years earlier, in 1932, Pfeiffer had discussed the desirability of arriving at an internationally standardised method of identification at the Zürich Congress of International Association of Testing Materials (Pfeiffer and Varossieau, 1945). Clarke's multiple-entry marginally perforated card system began a movement toward standardisation but many different cards were designed. Cards developed in one country were not compatible with cards developed elsewhere. Varossieau and Kukachka proposed a standard method which combined an 80-column card coupled with a card sorting machine and a standard list of characters arranged on the card in a particular way. The idea was a step forward but computers and card-sorting equipment were new and expensive and most wood anatomists did not have access to such equipment. Since 1950, multiple-entry marginally perforated cards and dichotomous keys dominated the field of wood identification and little, if anything, was accomplished on the automated punch-card system.

Non-statistical approaches for computerised wood identification are only beginning to emerge from the past decade's development of computer-assisted specimen identification in biology. A comprehensive review of the early work in this field is found in Morse (1974, 1975) and other papers in the Systematics Association Special Volume 7 edited by Pankhurst (1975). Of the computerised wood identification programs written, two were prepared under the direction of wood anatomists specifically for wood identification from data on marginally perforated punched cards (Pearson and Wheeler, 1981, and Hillis*, pers. comm.).

A third system under development (Miller, 1980) is based on a program written by L.E. Morse (Morse *et al.*, 1971; Morse, 1974) for the identification of biological specimens. This program incorporates many options that can assist anatomists in making a positive identification in a short period of time and without a great deal of experience.

A new approach to wood identification based on stereological techniques and computer manipulation of statistical data was proposed by Steele *et al.* (1975, 1976). This method quantifies many of the subjective traditional wood identification characters. For this quantitative characterisation stereological countings are applied to transverse and tangential sections of some selected species of hardwoods that are closely related and widely different. Vessels, fibres, and ray cells together with their walls and lumina are recognised. Volume fractions, average size, average spacing, aspect ratios for oriented features, and size distribution parameters are determined from point and intercept counts. These raw data are then analysed with a standard multivariate discriminant analysis program. Each species, including those that are very similar, are identifiable by the quantified data in the program to determine which pattern most closely fits the unknown. This stereological approach can be used to identify woods from distinctly different trees but its greatest potential is its ability to separate closely related species that cannot otherwise be readily distinguished. Another and perhaps a more practical use for the quantitative characterisation of wood is the prediction of properties and utilisation potential of various tropical species.

Computerised wood identification is another tool to be added to present-day techniques. Today's methods of wood identification are confined to immediate recognition, expert identification, dichotomous keys, and marginally perforated cards or tables and charts. The adoption of the computer for wood identification does not mean that the methods used today will disappear into obscurity. There will always be immediate recognition and expert identification of well known taxa. The dichotomous keys and marginally perforated cards have proven to be invaluable for routine work and small groups of taxa. In addition,

---

* Dr. W.E. Hillis, CSIRO Division of Building Research, Forest Products Laboratory, Highett, Victoria, Australia.

dichotomous keys are easily circulated via the printed page and thus are more readily available and cheaper to use. In this respect there are many computer programs, including the one written by Morse (1974), designed to assist the scientists in developing dichotomous keys.

What computerised wood identification offers wood anatomists is a data base that can be readily searched for difficult samples that cannot be identified through the use of standard techniques, *e.g.* samples of which origins are unknown or where the answers obtained from dichotomous keys do not compare with unknowns. It also offers wood anatomists with little experience in identification the opportunity to attempt an identification of a difficult sample. Of course, with more experience in classical identification the more likely a positive identification will be made. The computer, however, enables the inexperienced wood anatomist to become proficient in a much shorter time and for a routine identification service it offers speed and low cost.

In addition to the computer's ability to search and manipulate hundreds of species or taxa with great speed and accuracy, other advantages include the multiple-entry and polythetic capabilities. Dichotomous keys dictate the order of character selection, which often causes problems when the character dictated by the key is absent or obscure. Multiple-entry keys such as the marginally perforated cards and the computer allow the user to select characters that he wishes. The polythetic capability, which is only possible with the computer, is a powerful option. In the monothetic mode, a taxon is eliminated from the remaining possibilities when one difference between the taxon description and the unknown specimen is found. In the polythetic mode, a taxon is eliminated when more than one difference is accumulated. In this way an atypical specimen may be correctly identified even though some characters may have been misinterpreted. Further the polythetic mode has the advantage of allowing a considerable amount of intra-specific variation.

The biggest problem with computerised wood identification falls not with the ability of the computer or the program but with gathering and organising data into a standard format for the computer. There is an abundance of wood anatomical data available in the literature already on marginally perforated cards but there was no standard format or list of characters with which this mass of data could be converted into a readable format for the computer.

A Committee of the International Association of Wood Anatomists, coordinated by R.B. Miller and P. Baas, has prepared a standard list of characters for computerised hardwood identification (IAWA, 1981) and data can now be organised on data sheets from literature sources, marginally perforated cards, or reliable wood specimens. When complete data are not available from the literature or marginally perforated cards, samples from wood collections will be examined to complete the data.

Certainly computerised wood identification is not a new concept, but it has come of age. The increased memory storage, speed and accesibility, and the decrease in cost in computer technology offers the opportunity to advance the methodology of wood identification from pre-World War II to the 1980s. Programs and systems for identification are written, documented, tested, and being successfully used in other fields. We, as wood anatomists, need only to organise the wood anatomical data to conform to a format for a particular program.

References

Anderson, T.W. 1958. Introduction to multivariate statistical analysis. John Wiley & Sons, New York.

Andrew, I.A. & J. Burley. 1972. Variation of wood quality of Pinus merkusii Jungh. and de Vriese; five trees of Burma provenance grown in Zambia. Rhod. J. Agric. Res. 10: 183-202.

Bailey, I.W. & H.B. Shephard. 1915. Sanio's laws for the variation in size of coniferous tracheids. Bot. Gaz. 60: 66-71.

Balan Menon, P.K. 1957. Microscopic key for the identification of commercial timbers. Malay Forester 20: 162-164.

Beversluis, J.R. 1943. De micrografische identificatie van conifere houtsoorten. Meded. Landb. Hogesch. Wageningen 47/2: 39 pp.

Bliss, C.I. 1967. Statistics in biology. McGraw-Hill, New York.

Brazier, J.D. & G.L. Franklin. 1961. Identification of hardwoods. For. Prod. Res. Bull. No. 46: 96 pp.

Burley, J. 1969. Tracheid length variation in a single tree of Pinus kesiya Royle ex Gord. Wood Sci. Technol. 3: 109-116.

Burley, J. 1970. Variation in wood properties of Pinus kesiya Royle ex Gordon (syn. P. khasya Royle; P. insularis Endlicher): eighteen trees of Burma provenance grown in Zambia. Wood Sci. Technol. 4: 255-266.

Burley, J. & I.A. Andrew. 1970. Variation in wood properties of Pinus kesiya Royle ex Gordon (syn. P. khasya Royle; P. insularis Endlicher): six trees of Assam provenance grown in Zambia. Wood Sci. Technol. 4: 195-212.

Burley, J. & P.M. Burrows. 1972. Multivariate analysis of provenance variation in needles of Pinus kesiya Royle ex Gordon (syn. P. khasya Royle; P. insularis Endlicher). Silvae Genetica 21: 69-77.

Burley, J., T. Posner & P. Waters. 1970. Sampling techniques for measurement of fibre length in Eucalyptus species. Wood Sci. Technol. 4: 240-245.

Cailliez, F. & P. Gueneau. 1972. Principal component analysis of technological properties of Malaysian timbers. Cahier Scientifique No. 2, Suppl. Bois For. Trop.: 51 pp.; and Ann. Sci. For. 29: 215-265.

Clarke, S.H. 1938. A multiple-entry perforated-card key with special reference to the identification of hardwoods. New Phytol. 38: 369-374.

Corothie, H. 1948. Maderas de Venezuela. Min. Agric. y Cria, Caracas.

Cutler, D.F. 1979. The scanning electron microscope in recent systematic plant anatomy. In C.R. Metcalfe & L. Chalk (eds.), Anatomy of the Dicotyledons. Vol. 1 (2nd Ed.). Oxford Univ. Press, Oxford.

Dadswell, H.E. & A.M. Eckersley. 1941. The card sorting method applied to the identification of the commercial timbers of the genus Eucalyptus. J. Counc. Sci. & Ind. Res. 14: 266-280.

Dancik, B.P. & B.V. Barnes. 1975. Leaf variability in yellow birch in relation to environment. Canad. J. For. Res. 5: 395-399.

Davidson, J. 1972. Variation, association and inheritance in Eucalyptus deglupta Blume. Ph.D. Thesis. Austr. National Univ., Canberra.

Falkenhagen, E.R. & N. Nash. 1978. Multivariate classification in provenance research. Silvae Genetica 27: 14-23.

Freese, F. 1967. Elementary statistical methods for foresters. U.S.D.A. Agric. Handbook 317: 87 pp.

Gottwald, H. 1958. Handelshölzer. Ihre Benennung, Bestimmung und Beschreibung. F. Holzmann, Hamburg.

Gower, J.C. 1975. Relating classification to identification. In: R.J. Pankhurst (ed.), Biological Identification with Computers: 251-263. Acad. Press. London.

Graaff, N.A. van der, & P. Baas. 1974. Wood anatomical variation in relation to latitude and altitude. Blumea 22: 101-121.

Hans, A.S. & J. Burley. 1972. Wood quality of eight Eucalyptus species in Zambia. Experientia 28: 1378-1380.

Hans, A.S., J. Burley & J.G. Williamson. 1972a. Wood quality of Eucalyptus grandis (Hill) Maiden in a fertilizer trial at Siamambo, Zambia. E. Afr. Agric. For. J. 38: 157-161.

Hans, A.S., J. Burley & P. Williamson. 1972b. Wood quality in Eucalyptus grandis (Hill) Maiden, grown in Zambia. Holzforschung 26: 138-141.

Hogeweg, P. & J. Koek-Noorman. 1975. Wood anatomical classification using iterative character weighing. Acta Bot. Neerl. 24: 269-283.

Hopp, H. 1941. Methods of distinguishing between ship-mast and common forms of black locust on Long Island. N.Y. USDA Techn. Bull. 742: 24 pp.

IAWA Committee. 1981. Standard list of characters suitable for computerized hardwood identification. IAWA Bull. n.s. 2: 99-145 (also separately: IAWA, Leiden).

Jeffers, J.N.R. 1959. Experimental design and analysis in forest research. Almqvist & Wiksell, Stockholm.

Jeffers, J.N.R. 1962. Principal components analysis of designed experiments. The Statistician 12: 230-241.

Jeffers, J.N.R. 1967a. The study of variation in taxonomic research. The Statistician 17: 29-43.

Jeffers, J.N.R. 1967b. Two case studies in the application of PCA. Appl. Stat. 16: 225-236.

Kendall, M.G. 1975. Multivariate analysis. Charles Griffin, London.

Kettenring, J.R. 1971. Canonical analysis of several sets of variables. Biometrika 58: 433-451.

Koek-Noorman, J. & P. Hogeweg. 1974. The wood anatomy of Vanguerieae, Cinchoneae, Condamineae, and Rondeletieae (Rubiaceae). Acta Bot. Neerl. 23: 627-653.

Kribs, D.A. 1959. Commercial foreign woods on the American market. Pennsylvania State Univ.

Kukachka, B.F. 1960. Identification of coniferous woods. TAPPI 43: 887-896.

Mariaux, A. & A. Vitalis-Brun. 1979. A tentative statistical method for wood identification in 20 genera of Sapotaceae. IAWA Bull. 1979/2: 39.

Mergen, F., D.T. Lester, G.M. Furnival & J. Burley. 1966. Discriminant analysis of Eucalyptus cinerea x Eucalyptus maculosa hybrids. Silvae Genetica 15: 148-154.

Metcalfe, C.R. & L. Chalk. 1950. Anatomy of the Dicotyledons. Vol. 1 & 2. Clarendon Press, Oxford.

Metcalfe, C.R. & L. Chalk. 1979. Anatomy of the Dicotyledons. Vol. 1 (2nd Ed.). Oxford Univ. Press, Oxford.

Metcalfe, C.R. & L. Chalk. 1982. Anatomy of the Dicotyledons. Vol. 2 (2nd Ed.). Oxford Univ. Press, Oxford.

Miller, R.B. 1980. Wood identification via computer. IAWA Bull. n.s. 1: 154-160.

Morse, L.E. 1974. Computer programs for specimen identification, key construction and description printing using taxonomic data matrices. Michigan State Univ., Publ. Mus. Biol. Ser. 5: 1-128.

Morse, L.E. 1975. Recent advances in the theory and practice of biological specimen identification. In: R.J. Pankhurst (ed.), Biological Identification with Computers. Acad. Press, London.

Morse, L.E., J.A. Peters & P.B. Hamel. 1971. A general data format for summarising taxonomic information. Bioscience 21: 174-180, 186.

Normand, D. 1946. Les clefs pour l'identification des bois et le système des fiches perforées. L'Agron. Trop. 1: 162-172.

Normand, D. 1972. Manuel d'identification des bois commerciaux. I. Généralités. Centre Techn. For. Trop., Nogent-sur-Marne, France.

Oever, L. van den, P. Baas & M. Zandee. 1981. Comparative wood anatomy of Symplocos and latitude and altitude of provenance. IAWA Bull. n.s. 2: 3-24.

Pankhurst, R.J. 1975. Biological identification with computers. Syst. Assoc. Spec. Vol. 7. Acad. Press, London.

Pearson, R.G. & E.A. Wheeler. 1981. Computer identification of hardwood species. IAWA Bull. n.s. 2: 37-40.

Pfeiffer, J.P. & W.W. Varossieau. 1945. Classification of the structural elements of the secondary wood of dicotyledons using decimal indices for classification and identification of wood species. Blumea 5: 437-489.

Phillips, E.W.J. 1941. The identification of coniferous woods by their microscopic structure. J. Linn. Soc. 52: 259-320.

Reyment, R.A. 1969. Biometrical techniques in systematics. Systematic Biology, Proc. Intern. Conf., National Acad. Sci. Washington, D.C.; repr. Publ. Palaeo. Inst. Univ. Uppsala No. 87: 542-587.

Ross, J.H. & J.W. Morris. 1971. Principal components analysis of Acacia burkei and A. nigrescens in Natal. Bothalia 10: 437-450.

Rouvier, R. 1966. L'analyse en composantes principales: son utilisation en génétique et ses rapport avec l'analyse discriminatoire. Biometrics 22: 343-357.

Sanio, K. 1872. Über die Grösse der Holzzellen bei der gemeinen Kiefer (Pinus silvestris). Jahrb. Wiss. Bot. 8: 401-420.

Scheffé, H. 1959. The analysis of variance. John Wiley & Sons, New York.

Seal, H.L. 1964. Multivariate statistical analysis for biologists. Methuen & Co., London.

Snedecor, G.W. & W.G. Cochran. 1967. Statistical methods. 6th Ed. Iowa State Univ. Press, Ames, Iowa.

Sokal, R.R. & P.H.A. Sneath. 1963. Principles of numerical taxonomy. Freeman, San Francisco.

Steele, J.H., G. Ifju & J.A. Johnson. 1975. Application of stereological techniques to the quantitative characterization of wood microstructure. Proc. 4th Intern. Congr. Stereology, Nat. Bur. Standards Special Publ. 431: 245-256.

Steele, J.H., G. Ifju & J.A. Johnson. 1976. Quantitative characterization of wood microstructure. J. Microscopy 107: 297-311.

Sudo, S. 1963. Identification of tropical woods. Bull. Govt. For. Exp. Sta. Meguro 157: 1-262.

Thunell, B. & E. Perem. 1947. Identifiering av träslag. Meddelanden Svenska Träforskning institutet. Trätekniska Avdelning 12: 1-12.

Varossieau, W.W. 1948. The identification of wood species with the aid of the Hollerith system. Blumea 6: 229-242.

Varossieau, W.W. & B.F. Kukachka. 1950. Suggestions for a standardized method of identification with mechanically punched cards. Proc. IAWA Meeting, Stockholm: 1-5.

Zobel, B.J., R.C. Kellison, M.F. Matthias & W.V. Hatcher. 1972. Wood density of the southern pines. North Carolina Agric. Exp. Sta. Techn. Bull. No. 208: 56 pp.

Index*

---

* Names of persons have only been included if referred to in a historical context.
  References to illustrations are marked with an asterisk.

244

248